T0235723

History of Analytic Philosophy

Series Editor: **Michael Beaney**

Giusseppina D'Oro
REASONS AND CAUSES
Causalism and Non-Causalism in the Philosophy of Action

Anssi Korhonen
LOGIC AS UNIVERSAL SCIENCE
Russell's Early Logicism and its Philosophical Context

Franz Prihonsky
THE NEW ANTI-KANT (*trans.* Sandra Lapointe)

Consuelo Preti
THE METAPHYSICAL BASIS OF ETHICS
The Early Philosophical Development of G.E. Moore

Erich Reck (*editor*)
THE HISTORIC TURN IN ANALYTIC PHILOSOPHY

Maria van der Schaar
G.F. STOUT
The Psychological origin of Analytic Philosophy

HISTORY OF ANALYTIC PHILOSOPHY
Series Standing Order ISBN 978–0–230–55409–2 (hardback)
Series Standing Order ISBN 978–0–230–55410–8 (paperback)
(*outside North America only*)

You can receive future titles in this series as they are published by placing a standing order. Please contact your bookseller or, in case of difficulty, write to us at the address below with your name and address, the title of the series and one of the ISBNs quoted above.

Customer Services Department, Macmillan Distribution Ltd, Houndmills, Basingstoke, Hampshire RG21 6XS, England, UK

Russell's Unknown Logicism

A Study in the History and Philosophy of Mathematics

Sébastien Gandon
Professor of Philosophy, Clermont Université,
Université Blaise Pascal, Clermont-Ferrand, France

© Sébastien Gandon 2012

Foreword © Michael Beaney 2012

Softcover reprint of the hardcover 1st edition 2012   978-0-230-57699-5

All rights reserved. No reproduction, copy or transmission of this publication may be made without written permission.

No portion of this publication may be reproduced, copied or transmitted save with written permission or in accordance with the provisions of the Copyright, Designs and Patents Act 1988, or under the terms of any licence permitting limited copying issued by the Copyright Licensing Agency, Saffron House, 6–10 Kirby Street, London EC1N 8TS.

Any person who does any unauthorized act in relation to this publication may be liable to criminal prosecution and civil claims for damages.

The author has asserted his right to be identified as the author of this work in accordance with the Copyright, Designs and Patents Act 1988.

First published 2012 by
PALGRAVE MACMILLAN

Palgrave Macmillan in the UK is an imprint of Macmillan Publishers Limited, registered in England, company number 785998, of Houndmills, Basingstoke, Hampshire RG21 6XS.

Palgrave Macmillan in the US is a division of St Martin's Press LLC, 175 Fifth Avenue, New York, NY 10010.

Palgrave Macmillan is the global academic imprint of the above companies and has companies and representatives throughout the world.

Palgrave® and Macmillan® are registered trademarks in the United States, the United Kingdom, Europe and other countries.

ISBN 978-1-349-36683-5   ISBN 978-1-137-02465-7 (eBook)
DOI 10.1057/9781137024657

This book is printed on paper suitable for recycling and made from fully managed and sustained forest sources. Logging, pulping and manufacturing processes are expected to conform to the environmental regulations of the country of origin.

A catalogue record for this book is available from the British Library.

A catalog record for this book is available from the Library of Congress.

10   9   8   7   6   5   4   3   2   1
21 20 19 18 17 16 15 14 13 12

Transferred to Digital Printing in 2013

Contents

List of Figures and Tables

Figures

Tables

Acknowledgements

Some of the ideas expressed in this book I have already expounded in print in different papers. In particular, in writing Chapters 1, 2 and 3, I have freely drawn upon four of my articles: 'Russell et l'*Universal Algebra* de Whitehead: la géométrie projective entre ordre et incidence (1898–1903)', *Revue d'histoire des mathématiques*, 10, pp. 187–256 (2004); 'Which Arithmeticisation for which Logicism? Russell on Quantities and Relations', *History and Philosophy of Logic*, 29(1), pp. 1–30 (2008); 'Toward a Topic-specific Logicism? Russell's Theory of Geometry in the *Principles of Mathematics*', *Philosophia Mathematica*, 17(1), pp. 35–72 (2009); 'Logicism and Mathematical Practices: Russell's Theory of Metrical Geometry in *The Principles of Mathematics* (1903)', in C. Alvarez and A. Arana (eds), *Analytic Philosophy and the Foundations of Mathematics*, Basingstoke: Palgrave Macmillan, forthcoming.

As I found it necessary to undertake a complete revision of the ideas presented in these papers, the reader should be warned that they do not correspond neatly with the chapters of the present book. I am grateful to the Société Mathématique de France, the Taylor & Francis Group, Oxford University Press and Palgrave Macmillan for having allowed me to use this material.

Several pages from Chapter 5 are reprinted from an article which will be included in the volume to be edited by N. Griffin and B. Linsky for the centenary of *Principia Mathematica*, and I am grateful to them for permission to reproduce this material.

The intellectual debts I incurred in writing this book are many and varied. While studying for my PhD (on Wittgenstein's *Tractatus Logico-Philosophicus*), I benefited from the advice of Christiane Chauviré and François Schmitz. During this period, several discussions I had with Jerôme Sackur on Wittgenstein, Russell and Frege proved indispensable – I owe him no less than my way of viewing Russell. More recently, my work on the history of analytic philosophy has benefited from interactions with many different people. I am especially indebted for comments, conversations and criticism to Patricia Blanchette, Nicholas Griffin, Gregory Landini, James Levine, Igor Ly, Mathieu Marion and Jean-Philippe Narboux.

Having obtained my PhD, I began exploring the philosophy and history of mathematics. I have greatly benefited from the intense activity one finds in Paris, and particularly from the friendly reception my work received at the REHSEIS research centre. I particularly appreciate the unique mixture of philosophers and historians of mathematics one encounters in this lab. I would like to thank Karine Chemla, Philippe Nabonnand, Marco Panza,

David Rabouin and Ivahn Smadja for having introduced me to the delights of the history and philosophy of mathematics.

My recent works in philosophy (and, in particular, this book) have been markedly influenced by the numerous discussions I have had with Brice Halimi. Certain questions that Brice raised with the most innocent expression on his face kept me awake at night. I hope in the near future to be able to return the favour.

I have the good fortune to be a member of the supportive department of philosophy at the Université Blaise Pascal in Clermont-Ferrand, and also a member of the PHIER (the research centre associated to it). I am particularly grateful to my colleagues Emmanuel Cattin, Laurent Jaffro and Elisabeth Schwartz for having created in Clermont a place open to philosophical research, which combines the French tradition in the history of philosophy with a deep interest in analytic philosophy and the philosophy of science.

For the last four years, I have, with Sébastien Maronne, co-organized a seminar in the Maison des Sciences de l'Homme on philosophy and the history of mathematics. This has been a further opportunity for me to interact with historians and mathematicians. I am also indebted to the Institut Universitaire de France, by which I was accepted as a junior member in 2008 and without whose support I would have been unable to commit the time involved in writing this book.

SÉBASTIEN GANDON

Series Editor's Foreword

During the first half of the twentieth century, analytic philosophy gradually established itself as the dominant tradition in the English-speaking world, and over the last few decades it has taken firm root in many other parts of the world. There has been increasing debate over just what 'analytic philosophy' means, as the movement has ramified into the complex tradition that we know today, but the influence of the concerns, ideas and methods of early analytic philosophy on contemporary thought is indisputable. All this has led to greater self-consciousness among analytic philosophers about the nature and origins of their tradition, and scholarly interest in its historical development and philosophical foundations has blossomed in recent years, with the result that history of analytic philosophy is now recognized as a major field of philosophy in its own right.

The main aim of the series in which the present book appears, the first series of its kind, is to create a venue for work on the history of analytic philosophy, consolidating the area as a major field of philosophy and promoting further research and debate. The 'history of analytic philosophy' is understood broadly, as covering the period from the last three decades of the nineteenth century to the start of the twenty-first century, beginning with the work of Frege, Russell, Moore and Wittgenstein, who are generally regarded as its main founders, and the influences upon them, and going right up to the most recent developments. In allowing the 'history' to extend to the present, the aim is to encourage engagement with contemporary debates in philosophy, for example, in showing how the concerns of early analytic philosophy relate to current concerns. In focusing on analytic philosophy, the aim is not to exclude comparisons with other – earlier or contemporary – traditions, or consideration of figures or themes that some might regard as marginal to the analytic tradition but which also throw light on analytic philosophy. Indeed, a further aim of the series is to deepen our understanding of the broader context in which analytic philosophy developed, by looking, for example, at the roots of analytic philosophy in neo-Kantianism or British idealism, or the connections between analytic philosophy and phenomenology, or discussing the work of philosophers who were important in the development of analytic philosophy but who are now often forgotten.

Bertrand Russell (1872–1970) is the most central figure of all in the emergence of analytic philosophy. His rebellion with Moore against British idealism at the turn of the twentieth century is one of the founding events of analytic philosophy. No less important was the logicist project that both

Frege and Russell pursued, in their different ways, in the thirty or so years around the turn of the twentieth century. Frege's aim was to show that arithmetic is reducible to logic, but his attempt foundered on the paradox that bears Russell's name. Russell's aim was to show that geometry as well as arithmetic is reducible to logic, and he had to find a way of solving the paradox as well as providing richer logical resources to deal with geometry.

Russell's first attempt was in *The Principles of Mathematics* of 1903, and his second was in *Principia Mathematica*, co-authored with Whitehead and published in three volumes in 1910, 1912 and 1913. These are the key works that Sébastien Gandon considers in the present book. The literature on these works is now extensive, but, as Gandon points out, most of the attention has focused on Russell's theory of logic and his philosophy of arithmetic rather than his theory of quantity and his philosophy of geometry. This is reflected in the topics that are familiar in the history of analytic philosophy: Russell's concern with relations and series, his 1903 conception of denoting, his 1905 theory of descriptions, his definition of the natural numbers as equivalence classes of classes, his theory of types, to name just some. According to Gandon, however, this leads to a distorted view of the kinds of analysis that are characteristic of Russell's work, and, given the centrality of that work, of analytic philosophy itself.

Focusing on the relatively unknown aspects of Russell's logicism, Gandon argues, enables us to explore an issue that raises profound questions concerning the nature of analysis: the relationship between pre-logicized and logicized mathematics, as Gandon puts it. The philosophical problem here is essentially the paradox of analysis: How can an analysis be both correct and informative? In the case of Russell's logicism, the problem can be stated as follows. If the logicist reconstruction captures everything that is present in pre-logicized mathematics, then how can it be informative? If, on the other hand, it reforms pre-logicized mathematics, then how can it be correct? Gandon identifies two extreme responses to the problem, which he calls the Fregean and the Wittgensteinian. On the former, the logicist is taken as not beholden to pre-logicized mathematics. On the latter, the working mathematician is allowed to reject logicized mathematics if it diverges. Whether or not these responses are justly described as Fregean and Wittgensteinian, Gandon is right to seek an intermediate path, and he builds a convincing case for interpreting Russell as offering such a path.

Gandon's detailed account of Russell's philosophy of geometry and theory of quantity provides the means to illuminate this path. In opposition to the first response, Gandon argues that pre-logicized mathematics provides constraints on logicist reconstruction, which need to be explained in the work of analysis. In opposition to the second response, Gandon argues that there is too much disagreement among working mathematicians for it to make sense to talk of capturing everything. Russell, according to Gandon, was not only sensitive to these disagreements but also sought to reconcile them in his logicist reconstruction through careful articulation. Logicism can itself contribute mathematically, clarifying and delineating mathematical content by integrating it in a logical framework. Logicization should be seen as a dynamic process, appreciated by recognizing all stages in the process, its value lying in the reflective equilibrium it establishes between the *ex ante* and *ex post* perspectives.

The Russellian solution to the paradox of analysis that Gandon develops in this book is, in my view, essentially right; it is also superbly illustrated in the book's own methodology. The textual and historical data are carefully selected, discussed and structured in a fruitful interpretive framework. It is both insightfully about, and an excellent exercise in, Russellian analysis.

MICHAEL BEANEY

Introduction

'These are some of the rules of classification and definition. But although nothing is more important in science than classifying and defining well, we need say no more about it here, because it depends much more on our knowledge of the subject matter being discussed than on the rules of logic.'

Antoine Arnauld and Pierre Nicole, *La logique ou l'art de penser* (II, xvi, 1683)

There is a large consensus in the literature that logical analysis is at the core of Bertrand Russell's thought.[1] The idea that doing philosophy is analysing knowledge remains a constant throughout the whole of Russell's career and plays a prominent role in the period which goes from the publication of *The Principles of Mathematics* (PoM) (1903) to the publication of the last part of *Principia Mathematica* (PM) (1913). Disagreements occur, however, when one gets to the stage of explaining what Russell's view of analysis was, and how it evolves. Thus, scholars disagree on the nature of certain particular analyses – for instance, on the interpretation of Russell's 1903 theory of denotation and proposition,[2] on the meaning of his 1905 doctrine of definite description[3] and on the reasons which led him to abandon his multiple relation theory of judgement in 1913.[4] They also disagree on how to answer certain general questions like: What are the common features of Russell's different analyses?[5] What is the ontological import of an analysis?[6] How is the relation between understanding and analysing a concept or a proposition to be conceived?[7] There is no doubt that the decomposition of complex terms in PoM, the analysis of definite descriptions and the late theory of judgement in PM are important examples of analyses. There is also no doubt that the questions concerning the unity of the analytic method, its ontological import and its epistemological significance are philosophically central. Though this book is also devoted to Russell's concept of analysis, I will not, however, address any of these issues. I will not discuss Russell's theory of denoting concepts, nor will I say much about incomplete symbols; and I will remain completely silent on the theory of judgement. Furthermore, I will not attempt to delineate the common core of Russell's various analyses; I will not take a stand on whether or not one should extract some onto-logical lesson from an analysis; and I will not say much on the principle of acquaintance. But, if I do not address what is unanimously considered to be the central issues in the field, what am I going to write about in this book?

Certain parts of Russell's works have been studied in depth. But others have been completely neglected. The 'remote' parts of PoM and PM, that is, the parts where Russell dealt with advanced mathematics, are still terra incognita, unexplored in the Russellian literature. In particular, Russell's theory of geometry (PoM VI) and his doctrine of quantity (PoM III and PM VI) have not yet been the subject of any substantial work.[8] I first set about writing this book to fill this gap.[9] Chapters 1 to 3 are thus devoted to Russell's theory of geometry in PoM VI and Chapters 4 to 6 are dedicated to his theory of quantity in PoM III and PM VI. But the impetus to write this book was not purely scholarly: it was to shed new light on Russell's logicism taken as a whole that I decided to embark on the venture. Indeed, the doctrines of space and quantity are not merely new examples of analyses: they are new kinds of examples, in the sense that they display some features which are lacking, or are more difficult to see, in the other more well-known cases. I do not claim that these examples are intrinsically more important than the ones usually considered in the literature. My suggestion is rather that scholars, when dealing with Russell's conception of analysis, nourish their thinking with only one kind of example and then become the victims of a one-sided diet.[10] By introducing new examples, my hope is to enrich the reflection on Russell's concept of analysis – to raise new questions, to smoke out new problems. In which respects then do the analyses of space and quantity exemplify a new kind of analysis? What are the features which are displayed in those cases but which are lacking in the better-known ones?

Russell decomposed his philosophical task into two different components (PoM, p. xx):

> The present work has two main objects. One of these, the proof that all pure mathematics deals exclusively with concepts definable in terms of a very small number of fundamental logical concepts, and that all its propositions are deducible from a very small number of fundamental logical principles, is undertaken in Parts II–VII ... The other object of this work, which occupies Part I, is the explanation of the fundamental concepts which mathematics accepts as indefinable ... The discussion of indefinables ... forms the chief part of philosophical logic.

The cases scholars usually focus on are the analyses presented in PoM I (the theory of the variable, of the class, of the proposition, etc., which all belonged to philosophical logic). The analyses of the mathematical concepts, expounded in the parts which follow, are usually neglected (I will mention the development of the finite cardinal below). Now, in philosophical logic, the concepts analysed often did not have a strictly fixed meaning before the emergence of the new logic in the works of Russell. Take the case of the notion of the denoting concept: Russell coined the term, and almost nowhere is there a discussion of this concept before PoM. Or take the example of the analysis of a class: because of Russell's paradox, the content of the notion was, at the beginning of the twentieth century, heavily debated, and part of what Russell aimed at was to find an appropriate way of recasting the notion. Denotation, variable, class, proposition, etc. were the key concepts of a new science Russell was building at the time. He had to fix the meaning of these notions in such a way that: (1) mathematics could be deduced from them; (2) the paradoxes could be avoided; (3) a philosophical story could justify the basic definitions. The difficulty was huge, but, on one point, Russell had complete freedom of action: he was not constrained by the meaning that the scientists and philosophers gave to these concepts before him. Philosophical logic was Russell's (and Frege's) creation.

Compare now the cases of logical concepts with the cases of mathematical concepts like number, continuity, space and quantity. These notions had a fixed meaning that pre-dated PoM and PM and which Russell could not change. This sets a new constraint on the analysis – the *analysans* should be faithful to the essential features of the *analysandum* as it was understood by pre-Russellian mathematicians. As Russell explained (PoM, p. 3): 'the definition professes to be, not an arbitrary decision to use a common word in an uncommon signification, but rather a precise analysis of the ideas which, more or less unconsciously, are implied in the ordinary employment of the term'.

The logical definitions of number, infinity and space should display 'the ideas which, more or less unconsciously, are implied in the ordinary employment of the terms'. Russell could not say the same for the concepts pertaining to philosophical logic, because the uses of the terms 'class', 'denoting concept' and even 'proposition' were not fixed (or at least not to this degree) before the analysis begins. When Russell turned to mathematics, the content to be discussed was no longer supplied by logic alone: it came from outside, from the existing mathematics – and this fact poses some specific problems. How much of the existing content should be taken into account in the analysis? Is the logical analysis supposed to provide at the end all that which is implied in the ordinary mathematical use of the term? Or is it supposed to reflect only one aspect of this use? And, if so, which aspect?

Let me take a familiar example. Since Descartes, we know that every sentence of Euclidean geometry can be translated into algebra in such a way that if p is a theorem, then the translation of p is true.[11] One could thus say that the reduction of Euclidean geometry to algebra is formally correct, in the sense that it preserves the truth values of the propositions pertaining to Euclidean geometry. However, it would be odd to maintain that the algebraic definition of Euclidean space displays 'the ideas which, more or less unconsciously, are implied in the ordinary employment of the term'. Many geometers, in the nineteenth century, while understanding that the Cartesian reduction was formally correct, refused to identify Euclidean space with a numerical manifold. They considered that some 'important' and 'essential' features of geometrical knowledge were not faithfully represented in the Cartesian approach. But what exactly is an 'important' property? How are we supposed to set the 'important' features of a given mathematical theory apart from the 'unimportant' ones? The notion of 'importance' is so elusive that it seems that the wish to go beyond formal correctness puts an end to any hope of characterizing in an objective way what will count as a satisfactory analysis of a piece of mathematical knowledge. In our example, how can we characterize in a clear and consensual way the important features of synthetic geometry that are lost in the algebraic rewriting?[12]

Russell claimed that mathematics was nothing other than logic. This was however just a claim: he never denied that there was an apparent heterogeneity between logical concepts and mathematical ones. It is this gap, only provisional for Russell, which is at the root of our difficulty. In philosophical logic, there is a complete homogeneity between the method of analysis and the material to which it is applied. Propositions, variables, denoting concepts and propositional functions are concepts which are intimately connected with what logical analysis is. Any definition of analysis refers to them. But there is no such homogeneity in the analysis of mathematical concepts: the method here is applied to a content, which is prima facie given to logic from outside. Of course, at the end, one learns that this separation is an illusion, that the mathematical concepts considered are reducible to logical notions. But this is not the situation we face at the beginning. And this initial discrepancy raises a new question: How should we view the articulation between the logical analysis and the various subject-matters it is supposed to be applied to? Should we see analysis as a well-defined method that can be applied to any mathematical topic in the same uniform way? Or, on the contrary, should we consider that the nature of the analytic process 'depends much more on our knowledge of the subject-matter being discussed than on the rules of logic' (to quote the passage at the beginning of this chapter)?

We might think initially that Russell embraced the first alternative, since, despite the evolution of his thoughts on the nature of the analysis during the period 1903–13, he never doubted that there is a method of logical analysis which could be applied to any kind of knowledge whatsoever. The reading of the 'remote' parts of PoM and PM teaches us however that Russell went very far in the direction of adjusting his analyses to the fine-grained features of the various topics he was discussing. The detailed study of his theory of space and quantity will show us that his analysis cannot be seen as a well worked general theory that can be mechanically applied to different fields in a uniform way. So, is there a way to recast the notion of logical analysis so as to make room for the idea that it can be adjusted to the local features of what it is applied to?

A serious obstacle stands in the way of such a revision. Russell was a logical universalist.[13] Logic, for him, had no presupposition and could not be based on the results of any particular non-logical knowledge. But claiming that the way one analyses mathematics depends on the topics we are analysing amounts to relativizing the logical analyses to something which is not logical – namely, the differences between mathematical topics as they are given in the present state of development of the mathematical sciences. And this seems to lead to a rejection, or at least to a weakening, of Russell's universalist ambition.

One can formulate the difficulty in another way: Who is supposed to assess the success of the logicist programme? Is it the 'working' mathematician, who, with his or her first-hand knowledge of mathematics, seems to be in the best position to compare the pre-logicized mathematics with its representation in the logical system? Or is it the logician, who could easily attribute any discrepancy between the pre-logicized and the logicized mathematics as a consequence of a mix-up between logic and psychology? There is a gap between mathematics and logic that even the most stubborn logicist should recognize. At the end of the journey, mathematics will appear as a province of logic. But this is not so at the beginning. This book is about that hiatus. Which status should we grant it? Is the logical perspective we get at the end of the day, once the reduction is completed, the only valuable one? Or, on the contrary, should we demand that the final picture fits the target, as it appeared beforehand, from the perspective of the working mathematician?

This question is difficult. And it has no equivalent in the analyses pertaining to philosophical logic, where the target is not foreign to logic. I will tackle different forms of this problem at different places in this book: in Chapters 3 and 6, and more directly in the concluding Chapter 7. But let me make clear at the outset that the primary purpose of this book is not to settle the issue. It is to open it. The problems that are usually dealt with in the works about Russell's analysis (the problems of delineating its unity, of defining its ontological and epistemological significance) are both important and difficult. But they are not the only interesting ones. There is another issue which has not attracted much attention in the scholarship and which is yet crucial for our understanding of Russell's logicism: the complicated relation between pre-logicized and logicized mathematics. The first goal of this book is to convince the reader that Russell's analyses of mathematical concepts pose specific and crucial problems that are not visible when one goes no further than the first parts of PoM and PM.

*   *   *

There is something odd in the picture I have drawn. I have said that Russellian scholars usually focus on the analyses of logical concepts (denoting concepts, variables, classes, propositions, etc.) rather than on the analyses of mathematical notions. But there is one exception: Russell's analysis of cardinal numbers in PoM II and in PM III. The case is discussed in depth in nearly all the books dealing with Russell. The argument that the usual discussion about analysis suffers from a one-sided diet then appears to be mistaken. Finite numbers obviously had a fixed meaning before PoM and PM, and the analysis of cardinal numbers does supply the very kind of example which I said was missing. Why then focus on the two exotic examples of space and quantity, rather than deal with the more prosaic and well-known arithmetical case? What is exemplified by the theories of space and quantity that is not exemplified by Russell's analysis of cardinal numbers?

Before answering this question, let me first explain why arithmetic is usually considered as central. The standard story has it that Russell's logicism extended the arithmetization of mathematics and succeeded in reducing mathematics to the theory of whole numbers. Weierstrass, Cantor and Dedekind are famous for having constructed, each in their own way, the real field and real analysis from elementary arithmetic. Russell, after Frege, would have added his contribution to the construction, in defining whole numbers as sets of similar sets and in basing arithmetic on logic and set theory. Such an account is provided by Russell himself (1919, p. 5): 'having reduced all traditional pure mathematics to the theory of the natural numbers, the next step in logical analysis was to reduce this theory itself to the smallest set of premises and undefined terms from which it could be derived'. If this interpretative framework were true, then scholars would be right to focus on arithmetic: the analysis of cardinal numbers would be the only place where mathematics and logic would interact. The other analyses (of space, quantity, etc.) would not teach us anything new, since they would just explain how these mathematical theories could be reduced to cardinal arithmetic, and not directly to logic. Thus, there is a tight connection between the idea that Russell's logicism is an extension of arithmetization and the importance scholars usually attach to the logical (set-theoretical) definition of cardinal numbers. But is it true that Russell adhered to the arithmetization programme?

There is no doubt that Russell was deeply influenced by the works of Cantor, Dedekind and Weierstrass.[14] Moreover it is a fact that his theory of real analysis and his theory of relation-number in PM IV are based on the works of Cantor and Dedekind. It is nevertheless a mistake to view Russell's logicism as a straightforward extension of the arithmetization programme. Russell rejected both the definition of geometrical space as a numerical manifold and Dedekind's cut-construction of real numbers.[15] On these two crucial points, Russell did not agree with the arithmetizers. The first interesting reason for studying Russell's theory of space and quantity is precisely to show that the widespread view that logicism is a continuation of arithmetization ought to be rejected. To convince the reader not to reject my unorthodox interpretation out of hand, let me just quote a letter from Whitehead to Russell (14 September 1909):[16]

> The importance of quantity grows upon further considerations – <u>The modern arithmeticisation of mathematics is an entire mistake</u> ... To consider [arithmetical entities] as the sole entities involves in fact complicated ideas by involving all sorts of irrelevancies – In short the old fashioned algebras which talked of 'quantities' were right, if they had only known what 'quantities' were – which they did not.

I will come at length to the content of this letter in Chapters 5 and 6. But note that it is Whitehead himself, who, in the passage just quoted, underlined 'the modern arithmeticisation of mathematics is an entire mistake'.

Now, if Russell did not endorse without reservation the arithmetization programme, then the privileged status attached to the construction of whole numbers in the literature could not stand unchallenged: why accord so much importance to a case (arithmetic) which should not be viewed as a central one? If Russell did not consider arithmetic as the narrow channel through which mathematics interacts with logic, then one can no longer ignore his other analyses. In order to get a clear view of the relations between logic and mathematics, one should also take into consideration his theory of space, quantity and real numbers.

There is, however, another reason to challenge the exclusive focus made in the literature on the arithmetical case. The analysis of cardinal numbers has some very special features which contribute to marking it off from other analyses. In PoM, Russell believed that his definition of cardinality in terms of classes of similar classes allowed him to prove all the Peano axioms. But the PoM approach did not resist the paradoxes, and Russell eventually came to realize that the existence of a class of individuals of any finite cardinality could not be derived from logical principles. Thus, in PM, Russell and Whitehead reluctantly[17] included an axiom of infinity in the list of the primitive propositions – an axiom whose status (is it a logical or an empirical truth?) is not completely clear. This story, which has been studied in detail,[18] exhibits, at a very general level, a quite simple pattern: Russell initially thought that a reductive path from arithmetic to logic could be found; but he finally came to realize that no such path existed and became resigned to enlarging the number of logical axioms. This shows that the main challenge that arithmetic raised for the logicist took the form of an existence issue: is there a way of reducing arithmetic to logic? No doubt, the challenge was very difficult. But it was not the only difficulty Russell was confronted with in his analyses of mathematical concepts. Beside the existence problem, he faced an issue of uniqueness. Take the example of projective space in PoM VI. Russell gave three definitions of the notion which, in the first, was characterized as a numerical manifold (by using the homogeneous coordinates); in the second, as an ordinal structure (as in Pasch, 1882) and, in the third, as an incidence structure (as in Pieri, 1898). From a 'formal' point of view, these different definitions were equally perfect: we could formulate them in purely logical terms, and all of them allowed us to derive the whole body of projective geometry. Yet, Russell did not content himself with embracing them all, or with choosing one at random. As we will see, he argued at length in favour of the third definition. Russell, in this case, as in many others,[19] used some non-formal criteria to select, among the different formally equivalent alternatives, one single analysis. The fact that Russell used some non-formal constraints to pick out one analysis is not visible in the cardinal case. In this case, he did not consider many different possible logical definitions, but tried hard to salvage the only one he knew.[20]

Let me summarize why I think that the arithmetic-centred interpretations contribute to hiding some interesting aspects of the articulation between logic and mathematics in PoM and PM. First, the focus on the definition of 'cardinal' is based on the idea that Russell followed the 'arithmetizers' in ascribing to the theory of numbers a special status within the mathematical sciences. This is false: the study of his 1903 theory of geometry and his 1913 theory of rational and real numbers leaves no doubt as to the fact that Russell aimed at opposing the arithmetical definition of space and real numbers. Arithmetic has no privileged status in his thought. There is thus no reason to favour the cardinal case over the definitions of series, space, real numbers, quantity, and so on. Second, the arithmetical case is a very particular one, in that the issue of the possibility of multiple reductions did not come up. This last issue is also central in the other analyses of mathematical concepts: Russell never stopped discussing and weighing up the various ways of characterizing the mathematical notions he considered. So, in addition to having no privileged status, the analysis of arithmetic has some peculiar properties that sets it apart from the others. An exclusive focus on this case is thus liable to distort Russell's views of analysis.

The reader will have, I hope, at the end of this book, a more precise idea of what is missed in the arithmetic-centred interpretation. But because the topic is important, I will try to be a little more precise. I have said that, unlike logical notions, mathematical concepts have a meaning which pre-dated the logical reconstruction. In fact, the issue is slightly more complicated. More often than not, mathematicians disagree about the content of their theories and concepts. What is given, before the analysis takes place, is thus not a clear-cut and well-defined body of knowledge (mathematics) – what are given are different ways of construing the architecture of this body, and also different ways of defining the basic mathematical concepts. Russell had to take into account not only the pre-logical meanings mathematicians gave to their concepts, but also the disagreements they had on this very question. The uniqueness issue I have just referred to is related to this problem: the various analyses among which Russell had to choose corresponded to different ways of viewing the global organization of the mathematical sciences. Let me take again the example of the alternative characterizations of projective space. The first definition of it as a numerical manifold was in line with the arithmetization programme, while the third one, which characterized it as an incidence structure, construed geometry as an independent branch of mathematics. So, in the first approach, arithmetic was seen as the basis of the whole of mathematics, while, in the third one, mathematics appeared more like an archipelago composed of different interconnected disciplines. At the time, the two conceptions were defended in the mathematical community and the question of the place of geometry within mathematics was highly disputed. This example shows then that the choice of a definition among many formally equivalent ones was not innocuous: it committed Russell to a certain view of mathematics and its internal structuring. Russell knew then very well that he could not rely on a pre-definite division of the mathematical field, and that the different possible reductions encapsulated deep disagreements on the manner of seeing mathematics as a whole.

The problem with the arithmetic based interpretation is that this conceptual horizon, made of conflicts and disagreements within the mathematical community, completely vanishes. In the standard approach, one imagines that Russell adhered to a view (the arithmetization view) that one assumes to have been espoused by all the mathematicians of the time. One then pretends that the real problem for Russell was to derive arithmetic from logic. As I said above, I do not contest the reality and the difficulty of this issue. My point is rather that this reconstruction, which conceals the existence of disagreements in the pre-logicized mathematics, leads us to ignore another problem, as daunting as the standard one: If different conflicting pictures of the global organization of mathematics are available, how can we determine the one that the logicist should follow? And how can we justify this choice? The peculiarity and difficulty of Russell's analyses of mathematical notions is nowhere more visible than when he dealt with the uniqueness problem. Thus, in order to justify his Pieri-inspired definition of projective space, Russell had to combine some very heterogeneous reasons, which came from logic but also from the history of geometry. This hybridization in reasoning has no equivalent in philosophical logic – it is what makes the analysis of mathematics specific. To really understand this idea of combining heterogeneous considerations, one should not only keep in mind that mathematical notions have a use before the logical analysis takes place, but also that this use is never conceptually transparent, that many different and conflicting ways of construing mathematics always coexist in mathematics.

I have not embarked on a detailed study of PoM VI and PM VI merely because I love all that is exotic and like to get my hands dirty. As a matter of fact, the issue of the relationship between pre-logicized and logicized mathematics does not emerge in all its intricacy and depth if we go no further than the analysis of the finite cardinals. Certain fascinating aspects of Russell's analytic practices simply cannot be observed when we focus on this case. This is why I decided to turn my attention to the remote parts of Russell's works – especially to his theory of geometrical space and quantity.

* * *

This book falls into three parts: Chapters 1 to 3 are devoted to Russell's theory of geometry in PoM; Chapters 4 to 6 are dedicated to his theory of quantity, first in PoM III (Chapter 4), then in PM VI (Chapters 5 and 6); Chapter 7 is of a more general tone, aimed at drawing some general conclusions from the detailed examinations developed in the first two parts. Chapter 1 is about Russell's theory of projective geometry. Russell made it clear that, for him, projective space was the fundamental geometrical structure. This thesis was not shared by everyone at the time and shows that his thought was rooted in a specific tradition which dated back at least to Von Staudt's pioneering works. In this chapter I will first present the contextual elements needed to understand Russell's reasoning. In particular, Klein has shown that Von Staudt's proof of the fundamental theorem of projective geometry was faulty, and a long discussion about the best way of repairing the proof grew from Klein's remarks. Some mathematicians thought that ordinal and metrical postulates were required; some believed that metrical considerations were needed; some, at last, didn't want to blur the purity of Von Staudt's construction and tried to derive the fundamental theorem within a strict Von Staudtian framework. The Italian mathematician M. Pieri followed this last path and succeeded at getting the fundamental theorem without introducing any new indefinable. In a second stage, I will insert Russell's theory of projective space into this intricate context. In PoM VI, Russell contrasted the ordinal definition of projective space inspired by Pasch (1882) to a purely projective characterization based on Pieri (1898). Russell clearly sided with Pieri: his theory of relations helped him to define in a precise way the nature of the projective method which Pieri continued to maintain.

In Chapter 2, I pursue the exploration of Russell's theory of space by expounding his view of metrical geometry. Starting from projective space and following the standard method elaborated by Klein (1871), Russell first showed how the notion of metric can be defined in terms of anharmonic ratio. But, in addition, he developed two other approaches: a direct axiomatic characterization of the concept and an approach where distance is seen as an empirical notion. As is well known, he finally chose to espouse this last view, that is, to consider metrical geometry as an empirical science, and distance as a non-logical indefinable. This is very surprising, since he had the technical means to reduce the notion of distance to logic (via the projective method or via a direct axiomatization). What were his reasons for expelling metrical geometry from the sphere of pure mathematics? I will attempt to answer this question by comparing Russell's theory to Poincaré's approach. In particular I will argue that, contrary to a very widespread belief, the two conceptions shared some common presuppositions, but that Russell and Poincaré diverged both on their ways of assessing the importance of projective geometry and on their theories of perception.

Chapter 3 is a first attempt to draw some preliminary conclusions from the detailed discussions developed in the first two chapters. In particular, I will criticize the if-thenist claim (brought forward in Musgrave 1977 and Coffa 1981) that any axiomatization of geometry mechanically leads to a logical reduction. This is obviously not the case, since Russell did not consider Pasch's analysis of projective geometry and Klein's projective derivation of metric as adequate analyses. He used some non-formal criteria to single out, among the many formally adequate possible analyses, the logically correct one. But the very use of these not merely formal criteria seems to threaten Russell's anti-psychologism, that is, the idea that we should not confuse the content of the mathematics with our knowledge of it. Indeed, the non-formal constraints which Russell referred to could always be suspected to result from a mix-up between the content of the theory and some extraneous elements. How then to be anti-psychologist while basing one's choice on certain criteria which could be easily exposed as proceeding from a psychological confusion? In Chapter 3 I will provide a first incomplete answer to this question by stressing the importance of the notion of a relational type. In PoM, Russell claimed that the main mathematical branches could be correlated to certain relational forms, called relational types. This correlation explains why he thought that some topic-specific (and thus some non-merely formal) features of the subject considered should be preserved in the analysis, without falling into psychologism (these topic-specific features were grounded on the logical relational types).

In Chapter 4, I expound the theory of quantity developed in the intricate PoM III. I first give a rough survey of the discussion revolving around the concept of magnitude at the beginning of the twentieth century in mathematics. Many different axiomatizations of quantity were developed, and this context is the background to Russell's thought. My second goal is to present the outlines of the PoM theory of quantity. Russell's original and systematic doctrine has, to my knowledge, never been studied in the literature, and this chapter is an attempt to fill this gap. My third aim is to explain why Russell espoused an absolutist view of magnitude when he endorsed a relativist conception of cardinals. This discrepancy is presented as a puzzle in the scholarship. I suggest that taking into account the specificities of his systematic doctrine, and in particular his division of the sphere of magnitude in non-overlapping kinds, helps us to solve this problem. Last, but not least, relying on the English original of *Sur la logique des relations* (left unpublished until its inclusion in volume 2 of the *Collected Papers*), I expound in some detail the relational theory of quantity Russell sketched out, but never developed, in PoM III. Such an exhumation is important because the relational doctrine of 'distance' is the first stage of the mature theory of vector family presented in PM VI. To study Russell's 1900 construction facilitates the grasp of his and Whitehead's more complicated 1913 theory.

Chapter 5 is a presentation of this latter doctrine, developed in PM VI, the last published part of the book. Russell and Whitehead's goal is to develop a theory of rational and real numbers which can account for the application of numbers in measurement (which satisfies the so-called 'application constraint'). In this chapter, I first expound the main outlines of the PM doctrine. I especially focus on the definition of numbers as relations of relations, and on the relational doctrine of vector family (which extended and generalized the former theory of 'distance'). I also draw a comparison between the PM doctrine and three other conceptions: Dedekind's approach, Burali-Forti's physicalist conception, and Frege's theory (as recently reformulated by Hale). Russell and Whitehead's solution is akin to Frege's one: in both theories, the real numbers are considered as logical entities (contra Burali-Forti) and are connected to quantities (contra Dedekind). But the way Russell and Whitehead implemented the application constraint was very different from Frege. To meet this constraint, Frege and Hale related the structure of numbers to the structure of quantitative domains. Russell and Whitehead's reasoning was typological rather than structural. In PM VI, numbers were made for measuring quantities because the types of numbers (relations of relations) were adjusted to the types of magnitudes (relations). This implementation of the application constraint sets Russell and Whitehead's doctrine apart from Frege and Hale's one – and I claim, in this chapter, that the former is conceptually much more flexible and stronger than the latter.

Chapter 6 extracts from the detailed discussion in the previous two chapters some general conclusions about the articulation between logic and mathematics in Russell's logicism. In particular, I focus on Russell and Whitehead's espousal of the application constraint. How can we justify it? Should we regard the constraint as a non-formal criterion allowing us to select the good definition among many possible ones? I will distinguish two conceptions of the application constraint. In the first, standard, conception, the constraint is seen as a substantive demand for connecting two bodies of knowledge (usually a mathematical theory and an empirical body of knowledge) which are conceived as distinct. In the second (which I label 'architectonic') account, the constraint is viewed as a tool for delineating the outlines of a concept or theory, whose content is not considered as already fixed. By analysing the way the theory of quantity and real numbers (PM VI) is connected to the theory of real analysis (PM V), I argue that Russell and Whitehead endorsed the weaker architectonic conception. In PM, the application constraint is not a demand aiming at connecting two already defined domains. On the contrary, the constraint is used (by the mathematicians themselves) only when there is no consensus as to the way we should carve out the mathematical field. The application principle is thus not in PM a general criterion, whose application would mechanically single out the good definition. It is more like a vague recommendation helping us to make definite the contested outlines of a body of knowledge – the application constraint is more like a rule of thumb, whose use should be determined on a case-by-case basis.

The lesson one can extract from the first six chapters is that, in his analysis of mathematics, Russell did not only aim at preserving theorem-hood (or truth), but also attempted to disclose the specificities of the main branches of existing mathematics. The study of what he did in the remote parts of PoM and PM leads us then to ascribe to him a topic-specific conception of analysis, according to which the criteria allowing us to single out the 'good' reduction depended on the subject considered. This confronts us with a serious problem, however. Claiming that, for Russell, the constraints put on the analysis varied from case to case presupposes that the distinctions between cases are given to us before the logical analysis takes place. Now, as I have already suggested, this directly clashes with Russell's universalist claim. For a universalist, only logical distinctions can be taken into consideration. The differences between the various existing mathematical branches are thus precisely the kinds of information that a logicist should ignore. How then can we reconcile Russell's universalism to his analytical practice, as it is exemplified in the remote parts of PoM and PM? How can we defend the idea that logic, as the first science, does not presuppose anything, and maintain at the same time that logical analysis should take into account the fine-grained differences between mathematical disciplines?

In Chapter 7, I have tried to find a way out of this dilemma. One could distinguish two non-Russellian ways of solving the question. The first, which I attribute to Frege, is to say that the logical definitions of the mathematical notions are all stipulations, and that therefore the logicist can and should completely ignore the issue concerning the fit between pre-logicized and logicized mathematics. The second, which I attribute to Wittgenstein, is to say that the working mathematicians have always the authority to assess the faithfulness of the logicist picture, and that, owing to the importance that notations have in mathematics, it is likely that the working mathematicians would reject all the translations offered by the logicist. Russell did not follow Frege – the remote parts of PoM and PM show that he discussed different ways of carving out the mathematical fields and took into account the disagreements among mathematicians. But neither did he follow Wittgenstein – he was a logicist, and thought that the content of mathematics is ultimately displayed in logical definitions. How then can we find a middle position between Frege and Wittgenstein?

At this stage, a distinction between two kinds of universalism, first proposed by Rivenc (1993), will help us. According to Rivenc, the chief difference between negative universalism (endorsed by Frege) and positive universalism (ascribed to Russell) has to do with the way the logical system is related to its non-logical context. The negative universalist draws a sharp separation between the logical system and existing mathematics, and regards any attempt to compare pre-logicized and logicized mathematics as nonsensical. The positive universalist, on the other hand, sees the articulation between the logical system and its surroundings in a more dynamical way: the demarcation line is for him a moving frontier, which evolves as the system shapes up. Thus, for Russell, what at first appears as something external to the logical system (a difference between two distinct mathematical topics, for instance) becomes later a part of it (the difference is expressed as a difference between two relational types). What is specific to the positive universalism is thus the importance it gives to progressivity and time. The right perspective is not, as in Frege, the ultimate one, the one that is obtained when the logical system is completed. The right perspective is the one arrived at when one considers the system in the making, when one circulates between the view one has *ex ante*, before the analysis takes place, and the view one has *ex post*, once the logical analysis has been achieved.

This opening of the system to its non-logical context should be related to the fact that, for Russell, the outlines of the mathematical content were not fixed once and for all in the 'real' mathematics. As I have already stressed, Russell was aware of the fact that his mathematician colleagues never stopped arguing about the best way of articulating their favourite theories. He knew well that there was no consensus, at the time, in existing mathematics as to the architecture of the mathematical field. This set his position apart from Wittgenstein's. To claim, as Wittgenstein did, that mathematicians always have the authority to assess the fit between pre-logicized and logicized mathematics is to downplay the importance of disagreements among mathematicians. With Russell, logicizing mathematics was precisely a way to give voice to the various opposing mathematical parties and to weigh up their different arguments so as to reach a considered agreement. As my study of PoM VI and PM VI show, Russell always developed different possible constructions of the concepts he was analysing and tried hard to integrate them in one coherent whole.

Russell's division of the mathematical material into distinct branches (that is, the organization of PM and PoM) was then not a mere copy of the organization of existing mathematics as it was conceived by the mathematicians. It was not yet considered a purely logical product, elaborated in an autarkic way, without any interaction with the pre-logicized mathematics. The organization of PoM and PM was justified in the course of their development by a constant back-and-forth process, by a reciprocal adjustment between the logical forms and the mathematical material. In Russell's brand of logicism, there was thus no opposition between the (*ex ante*) perspective of the working mathematicians and the (*ex post*) perspective of the logical foundationalist. The logical framework was rather for Russell a tool allowing us to explore and balance the pros and cons of the different ways of carving out mathematics, and to try and come to an agreement between the various opposing standpoints that the mathematicians developed. Our initial puzzle (how to reconcile a topic-specific conception of logical analysis with logical universalism) would then finally find a solution. Russell's universalism was of a dynamical kind. It was encapsulated in a progression more than in a system. The importance given to case-by-case analysis and the sensitiveness to topic-specific features would not be irreconcilable with the universalist ambition. On the contrary, all these characteristics would be a direct consequence of the kind of positive universalism Russell sought to develop.

The concluding chapter is of course much more general and speculative than the previous ones. Fortunately, nothing in what has gone before depends on what I present in the conclusion. As I said, my main purpose in this book is to raise a new question (or perhaps to re-articulate a forgotten one): How can we view the articulation of pre-logicized and logicized mathematics in Russell's thought? I do not pretend to have written the last word on this. There might be some other ways to explain and interpret what Russell did in PoM VI and PM VI, and I would be more than happy if future works could solve the philosophical questions I have attempted to articulate here.

1
Projective Geometry

In his first book, *An Essay on the Foundations of Geometry*, Russell, following Felix Klein,[1] explained how the three classical metrical three-dimensional geometries (the Euclidean, the hyperbolic and the elliptic) could be developed in a projective setting. The method was rather complicated (see Chapter 2), but the mathematical reasoning was widely known and used at the time. If Russell did not completely endorse the reduction of all the metrical concepts to the projective ones, he clearly assumed that metrical geometry presupposed the projective framework. For Russell, then, defining the nature of space amounted ultimately to defining the nature of projective space. He had not changed his mind by 1903. PoM VI is indeed composed of three parts: the first one (from chapters 44 to 46), devoted to the definition of space, is a study of projective geometry; the second part (from chapters 47 to 49) is dedicated to metrical geometry; the third one (from chapters 50 to 53) is more philosophical in nature – it is a rebuttal of certain arguments (from Lotze and Kant, aimed at showing that space is a contradictory concept).[2] There is thus no doubt that Russell, in PoM, resumed Arthur Cayley's slogan according to which 'projective geometry is all geometry'.

Two difficulties emerge, however, as soon as one enters more deeply into Russell's analysis of projective space. First, in chapter 44 (entitled 'Dimensions and Complex Numbers'), Russell defines space as a multiple series (that is, as a series with more than one dimension). Now it is not easy, at first, to understand the connection between this characterization and the presentation of projective geometry that follows. What is the relation between multiple series and projective space? Second, Russell gives two axiomatic characterizations of projective space: one in chapter 45 (entitled 'Projective Geometry') and another in chapter 46 (entitled 'Descriptive Geometry'). The former is inspired by the work of the Italian mathematician Mario Pieri,[3] the latter by the book of the German geometer Moritz Pasch.[4] It is difficult to understand why Russell placed side by side two different accounts of projective space. It is also difficult to see why he finally chose to favour the first one. In this chapter I will attempt to resolve these two interpretative problems, by showing that they are facets of the same coin. Indeed, for Russell, the crucial problem was to extract from projective geometry a convincing definition of space. Pieri's approach provided him with a way to do this, since Pieri had succeeded in presenting projective geometry as a theory of incidence relations, and thus as a theory of multiple series (that is, as a theory of series which can intersect themselves). Pasch's construal of projective geometry, based on order, even if it is formally correct, did not provide him with the same resource.

In this chapter, we will focus on Russell's definition of space and on his view of projective geometry – that is, our first task will be to expound Russell's two constructions of PoM 45 and PoM 46, which I will do in section 1.3, after having presented in sections 1.1 and 1.2 some contextual information about the state of projective geometry at the end of the nineteenth century. In section 1.4, I will explain why Russell sided with Pieri rather than Pasch. To do so, I will be compelled to come back to the geometrical parts of Whitehead's (1898) *A Treatise on Universal Algebra* and present Russell's little known criticisms of it. In section 1.5, I will return to chapter 44 and lay stress on the importance Russell attached to incidence relation in his definition of space. I will in particular show that Russell's insight found some echoes in the mathematical researches of the time.

1.1 Projective and metrical properties

Here I present some material which will be constantly used throughout this chapter and those that follow. The idea is to blend some mathematical explanations about the key principles of projective geometry with a selective presentation of its development during the nineteenth century. I say 'selective' because my aim is first and foremost to explain Russell's PoM VI – I will thus only focus on the features which played an important role in Russell's thought.[5]

Jean-Victor Poncelet (1788–1867) is the first to have viewed projective geometry as an independent branch of geometry, defined by its specific methods, its distinctive results and its particular problems. His key insight was to study, not (as in Euclid) the properties of one particular sort of figure (triangle, circle, etc.), but all the properties of a family of figures related by certain transformations.[6] Poncelet focused on a type of transformation called 'perspectivities'. Let O be a given point in Euclidean space, and call O-pencil the pencil of lines passing through O: the lines (OA), (OB) and (OC) of Figure 1.1 belong to the O-pencil. Let Π, Π' be two distinct planes not containing O. A set of Π'-points is said to be in perspective (from O) to a set of Π-points if and only if each point of the former set is related to a point of the latter by a line of the O-pencil. Thus, in Figure 1.1, the circle ABC is in perspective to the figure $A'B'C'$.

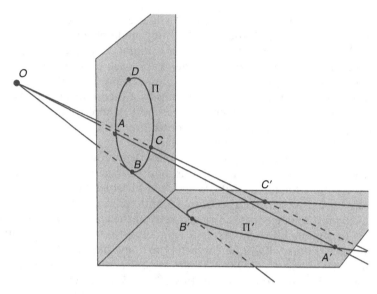

Figure 1.1 Two planar figures in perspective: a circle Π on the vertical plane is projected from O into a parabola Π', lying on the horizontal plane

This definition could be restricted to the two-dimensional case. Two ranges of collinear points are in perspective when they can be regarded as different sections of the same flat pencil. Thus, in Figure 1.2, the ranges of points Δ and Δ' are in perspective, because each element of the former can be related to (is a projection of) each element of the latter by a line of the pencil centred in O. The 'perspectivities' represent only one species of projective transformation, however. Poncelet defined the 'homographies' (his name for projective transformations) as a composition of perspectivities. That is, two planar figures (respectively, two ranges of collinear points) are projectively related if and only if they can be transformed one onto another by a series of perspectives.

At first sight, it seems that all the 'usual' properties of the figure are destroyed when a figure is projectively transformed. Thus, distances and angles are not invariant by projection (see the triangle ABC and its projection $A'B'C'$ in Figure 1.1). What is more the order of the points on a line is also not a projective invariant (in Figure 1.2, B is between A and C on Δ, while B' is not between A' and C' on Δ'). Nevertheless, some properties remain the same. In particular, incidence relations between the elements of a figure are usually (see below) not affected: lines are transformed into lines, pencils are transformed into pencils. Let's call 'projective' the properties that are invariant by homography, and 'metrical' those that are not. This distinction led Poncelet to differentiate between two kinds of theorem.

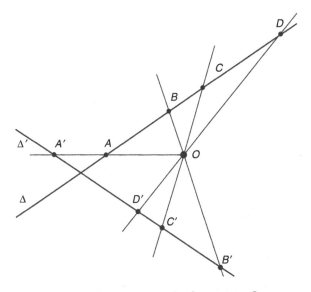

Figure 1.2 ABCD and A'B'C'D' are in perspective from centre O

Pythagoras's theorem is clearly a metrical theorem, because the property of being a right-angled triangle is not projective. On the other hand, the very important theorems of Desargues (see Figure 1.3) and Pappus (see Figure 1.4) are projective: they remain true when the figures they speak about are homographically transformed.

As Desargues and Pappus's theorems involved only incidence relations between points and lines in the projective plane, one could be tempted to believe that projective theorems are those which only deal with the incidence properties of the figures in the Euclidean plane. Things are more complicated, however. First, homographic transformations do not preserve all the Euclidean properties of incidence. Second, some projective properties seem to involve more than incidence. Let me explain these two important points in turn.

When projecting a given figure in the Euclidean space, some elements might vanish. Thus, in Figure 1.1, the point *D* on the plane *Π* will not have any correlate on *Π'*, because (*OD*) is parallel to *Π'*. To come back to the planar perspectivity illustrated in Figure 1.2, consider *E* on *Δ* such that *EO* is parallel to *Δ'*; *E* does not have any image on the Euclidean line *Δ'* by the perspectivity which transforms *AB* into *A'B'*. In order to make the transformation one–one, it is necessary to complete *Δ'* (and *Δ*) by introducing the point at infinity *E'*, the correlate of *E* by the perspectivity (*AB*, *A'B'*)[7] (see Figure 1.5). This introduction of the point at infinity 'closes' the line *Δ'* and makes it topologically akin to a circle.

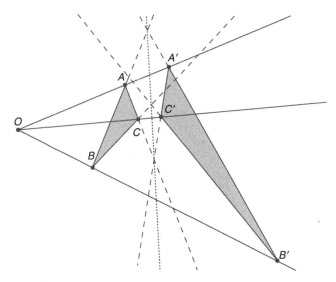

Figure 1.3 Desargues's theorem: the triangles *ABC* and *A'B'C'* on a projective plane are in perspective if and only if the intersections of the lines *AC* and *A'C'*, *BC* and *B'C'*, and *AB* and *A'B'* are collinear

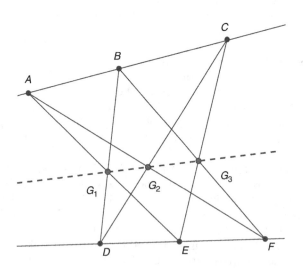

Figure 1.4 Pappus's theorem: let *A*, *B*, *C* be three points on a line and *D*, *E*, *F* be three points on another line of the projective plane. Let *BC* and *AE* meet at G_1, *CD* and *AF* meet at G_2, and *BF* and *EC* meet at G_3. Then G_1, G_2, G_3 are collinear

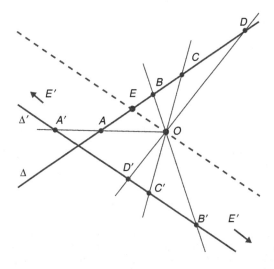

Figure 1.5 The image of *E* on *Δ'* by the perspectivity (*AB, A'B'*) is the 'point at infinity' *E'*

To ensure that incidence properties are preserved by projective transformation, one must then 'extend' the Euclidean space.[8] Today, the usual way to present this 'completion' is to define the projective plane as the set $\mathbf{P}^2$ of the vectorial lines of $\mathbb{R}^3$. An affine map of the projective plane is thus seen as a section of this set by an (affine) plane *Π* which does not pass through the origin *O*. Thus, in Figure 1.6, the affine map is represented by the shaded area parallel to the *x, y* plane. The line at infinity corresponds to the 'horizontal' vector lines belonging to the plane (*O, x, y*). Each line passing through *O* (and thus each projective point) can be represented by one of its vector **v**. To it, a triplet of real numbers (*a, b, c*) – not all equal to zero – can be associated. Mind however, that, as any multiple of *λ***v** (with *λ* real) can be associated with the same vectorial line, the coordinate of a projective point *P* is determined up to a common real factor. One refers to (*a, b, c*) – or as *λ*(*a, b, c*) – as the homogeneous coordinates of *P*. In Figure 1.6, the coordinate of the point at infinity *E* is (*e*, 0, 0), the one of the point at infinity *F* is (0, *f*, 0) – with *e* and *f* being real numbers.[9]

Poncelet laid great stress on this enrichment of the Euclidean plane, in which he saw the source of the fruitfulness of the new projective method. He maintained that points (and lines) at infinity enable the geometer to recover the strength of the algebraic methods (on the concurrence between analytic and synthetic method, see section 1.4). One example will suffice to grasp the idea. In analytic geometry, all the conic sections are represented by quadratic equations – while in Euclidean geometry, the three non-degenerate conic sections appear as divided into three different kinds: ellipse, hyperbola

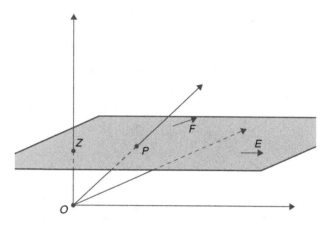

Figure 1.6   A representation of the real projective plane as the set of vectorial lines in $\mathbb{R}^3$. The shaded area is an affine map of the projective plane. Some points of the projective plane (like E and F) appear as points at infinity in the affine map. P has the coordinates (a, b, c), E $(e, 0, 0)$, F $(0, f, 0)$

and parabola. Now, by focusing attention on the equation, it is possible to extract several properties of the conics without bothering to distinguish between the different cases. The projective theory provided a new geometrical (i.e., non analytic) framework in which a unified treatment of conic sections could be developed. Indeed, in the projective plane, a conic section can be seen as the O-cone with axis (Oz) passing through the unit circle. Of this projective figure, various affine maps can be produced, depending on the relative situation of the cone and the intersecting plane. The distinction between hyperbola, ellipse and parabola are then not viewed as a difference between geometrical objects, but only as a difference between the different affine representations of the same projective entity.

Let me come back to the relation between projective property and incidence. Once Euclidean space is completed by the points and lines at infinity, the incidence relations between points, lines and planes become invariant by homography.[10] But one cannot yet identify the projective properties with the incidence properties, since, in the extended framework, projective transformations seem to preserve more than incidence. Let me take two examples, which will prove to have particular importance.

As we have seen, the projective line has the topological structure of the circle; and, in a circle, the betweenness relation does not have its 'expected' properties. For instance, if A, B, C are on a circle, then C can be between A and B, while B is between C and A: going clockwise from C, I pass through B on my way to A (see Figure 1.7). One can, however, replace the familiar

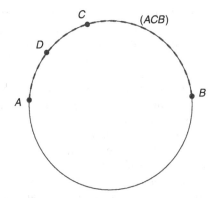

Figure 1.7 The segment (*ACB*) is the accentuated part of the circle

notion of betweenness by a four-term relation: one says that *A*, *C* separates *B*, *D* if one cannot go from *A* to *C* without crossing either *D* or *B*. One then defines the projective segment (*ACB*) as the set of points *X* such that *C* and *X* are not separated with respect to *A* and *B*.

Now, the interesting fact is that the relation of separation and the notion of 'projective segment' are projective invariants. Thus, in Figure 1.2, although the Euclidean order is not preserved by the transformation, the relation of separation between *AC* and *BD* is the same as the one between *A'C'* and *B'D'*. In other words, some ordinal relations between geometrical elements, which, prima facie, seem to have nothing to do with incidence, are projective.[11]

But there is more: some relations between distances and angles are invariant by projective transformations. Lazare Carnot (1753–1823) and August-Ferdinand Möbius (1790–1868) defined the cross-ratio between the two pairs of points *A*, *B* and *C*, *D* on the same line (noted (*AB*/*CD*)) as being equal to:

$$\frac{\overline{CA}}{\overline{CB}} : \frac{\overline{DA}}{\overline{DB}}$$

where $\overline{CA}$ is the algebraic distance between *C* and *A*.[12] When (*AB*/*CD*) = −1, one says that *A* is the harmonic conjugate of *B* with respect to *C* and *D*, or that *A*, *B*, *C*, *D* forms a harmonic set.[13] Now, while distances occur in its definition, a cross-ratio is a projective property of a set of four collinear points, that is, a quantity which is homographically invariant. In Figure 1.2 for instance, (*AB*/*CD*) is the same as (*A'B'*/*C'D'*). From a historical point of view, the concept of cross-ratio has played a crucial role in the emergence of projective geometry. For the French geometers (Carnot, Poncelet, Chasles), a cross-ratio was viewed as the basic projective concept, so much so that certain facts concerning alignment and incidence were derived from the invariance of the cross-ratios, and not the other way round.[14] Far from being viewed as the most fundamental ones, the incidence properties (the colinearity of P, Q and R) were thus derived from the equality between certain quantities. For the geometers of the first half of the nineteenth century, the fact that incidence properties were projective did not mean that projective geometry was about incidence – projective geometry was for them first and foremost the study of cross-ratios.

Let me pursue this development a bit further to anticipate the next section. We have seen that Poncelet defined a homography as a composition of perspectivities. The German mathematician Karl Georg Christian Von Staudt (1798–1867) developed another view, and characterized a projective transformation (in the real projective space) as a transformation which leaves the cross-ratio invariant. Von Staudt attempted to demonstrate the equivalence between his definition (in terms of cross-ratio invariance) and Poncelet's (in terms of the product of perspectivities). But for this, he needed the following result:

> Let ABC and $A'B'C'$ be two sets of points on two projective lines, then:
> (a) there is a projective transformation (a transformation which preserves cross-ratios) which maps A onto A', B onto B' and C onto C'.
> (b) there is only one such projective transformation.

This theorem became known as the fundamental theorem of projective geometry.[15] Von Staudt thought he had proved it, but Klein showed that his demonstration was flawed. As we will see below, this issue played a crucial role in Russell's thought.

But before pursuing this line, let me take stock of what we have seen so far. Projective geometry was seen as the study of projective properties of figures, that is, properties which were preserved by projective transformations. The properties related to the incidence between points, lines and planes (collinearity, coplanarity, etc.) were counted among these projective properties – at least as soon as Euclidean space was suitably completed. But other features were also seen as projective: the ordinal property of separation and the metrical notion of cross-ratio. In the mid-nineteenth century, it was thus not easy to characterize in a simple way the subject matter of projective geometry.[16]

1.2   The status of the fundamental theorem of projective geometry

By the time Russell became interested in the topic in the late nineteenth century, geometry had changed considerably. The most urgent task was no longer to challenge the Cartesian method, but to develop a new framework strong enough to make room for the new non-Euclidean geometries. But in a certain sense, the emergence of non-Euclidean spaces made the issue concerning the subject matter of the projective geometry more pressing. Indeed, as long as projective geometry was seen as a method for studying Euclidean figures, the question concerning the exact nature of the projective properties was marginal. But as soon as projective geometry was regarded as an independent theory, which could be put at the basis of the various metrical theories (Euclidean and non-Euclidean),[17] the issue became crucial – one could no longer avoid it by saying that the new method was only an auxiliary device to investigate the Euclidean world. Another important change (not unrelated to the first one, of course) is that geometers at the end of the nineteenth century started to axiomatize real projective geometry.[18] This motif was prominent in Russell, for chapters 45 and 46 of PoM present two axiomatic systems of real projective geometry. It would be mistaken however to attach to this new 'style' too much importance – projective geometers at that time (Hilbert included)[19] did not see axiomatization as an end in itself, but rather as a tool to explore and answer some issues already present in the works of Poncelet and his followers. In this section, I will relate the axiomatic treatment that blossomed from the 1880s onward to the difficulties surrounding the exact delimitation of the subject matter of projective geometry.

Before entering into the details, it is worth looking at a standard presentation. Let us take the classic axiomatic of real projective three-dimensional space given in Coxeter (1947, pp. 20–2). Here, one finds two groups of postulates. In the first there are seven axioms of incidence:

1. There are at least two points.
2. Any two points are incident with just one line.
3. The line AB is incident with at least one point besides A and B.
4. There is at least one point not incident with the line AB.
5. If A, B, C are three non-collinear points, and D is a point on BC distinct from B and C, while E is a point on CA distinct from C and A, then there is a point F on AB such that D, E, F are collinear.
6. There is at least one point not in the plane ABC.
7. Any two planes intersect in a line.

In the second there are seven axioms of order (the last one is singled out by Coxeter as an axiom of continuity):

8. If A, B, C are three collinear points, there is at least one point D such that AB separates CD.
9. If AB separates CD, then A, B, C, D are collinear and distinct.
10. If AB separates CD, then AB separates DC.

11. If *A*, *B*, *C*, *D* are four collinear points, then either *AB* separates *CD* or *AC* separates *BD* or *AD* separates *BC*.
12. If *AB* separates *CD* and *AD* separates *BX*, then *AB* separates *CX*.
13. If *AB* separates *CD* and *ABCD* are in perspective with *A'B'C'D'*, then *A'B'* separates *C'D'*. **Definition**: If *A*, *B*, *C* are three collinear points, we define the segment (*ABC*) as the class of points *X* for which *AB* does not separate *CX*.
14. For every partition of all the points of a segment into two non-vacuous sets, such that no point of either lies between two points of the other, there is a point of one set which lies between every other point of that set and every point of the other set.

In the standard axiomatic presentation given here, projective geometry appears then to be a combination of two subtheories: a theory of the incidence relations between points, straight lines and planes; and a theory of the ordinal relation of separation between pairs of points on a line.

With the incidence axioms only, many projective theorems can be proved. For instance, Desargues's theorem (Figure 1.3) is a direct consequence of the postulates of incidence.[20] More important and also more surprising, many projective concepts, at first sight not reducible to incidence, can be developed from the first seven axioms. The most notable case is that of the notion of the harmonic set. According to Von Staudt (1847), four collinear points *A*, *B*, *C*, *D* are said to form a harmonic set if there is a quadrilateral *EFGH*, such that the two opposite sides *GH* and *FE* pass through *A*, the two other *GF* and *HE* pass through *B*, while the diagonals *HF* and *GE* pass (respectively) through *C* and *D* (see Figure 1.8). From Desargues's theorem, it follows that if *A*, *B*, *C*, *D* forms a harmonic set, then any quadrilateral which is such that its two pairs of opposite sides intersect at *A* and *B*, and such that one of the diagonals passes through *C*, has the other diagonal which passes through *D*. This result means that to find the harmonic conjugate of *C* with respect to *A* and *B*, one only needs to construct a quadrilateral with the required properties and consider point *D*, which is the intersection of the line *AB* with the remaining diagonal. That the position of *D* thus obtained does not depend on the quadrilateral considered follows from the incidence axioms only.

To appraise the importance of this result, recall that, while considered as a projective property of a set of four collinear points, harmonic conjugation was defined, by Carnot, Poncelet and Möbius, as a ratio between ratios of distances or angles. This raised a difficulty: how to explain that something which was composed of non-projective notions (distance, angle) remained invariant by homography? Von Staudt's new approach accounted for this fact, since it replaced the 'metrical' characterization of harmonic conjugation by a (complicated) definition in terms of incidence relations in the plane. As we will see, Russell laid great stress on this definition.[21]

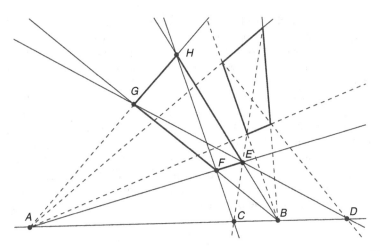

Figure 1.8 Von Staudt's quadrilateral construction: *B* is the intersection of *GF* and *HE*, *A* is the intersection of *GH* and *FE*, *C* of *AB* and *HF*, *D* of *AB* and *GE*. One can check that if one constructs another quadrilateral with exactly the same properties, then one ends up with exactly the same set of points

There are however some key projective theorems that cannot be proved without using the postulates belonging to the second group. This is notably the case of Pappus's theorem and of the fundamental theorem of projective geometry. As one can hardly overestimate the importance of this fact in the story I am about to tell, I will pause to explain it.

Recall that the fundamental theorem says that a projective transformation (a transformation which preserves cross-ratio) between two ranges of points is uniquely determined when we are given three points and their corresponding images. Von Staudt believed that an iteration of his quadrilateral construction could be used to prove this theorem, though this turned out to be false. Let me explain Von Staudt's reasoning. From three given points *A*, *B* and *C*, one can, using the quadrilateral construction, draw the three conjugates of each of the three points with respect to the two remaining ones; then, from the six points obtained, one can, by iterating the constructions, get six more points; and we can apply on the new set of points the same process again and again. The 'Möbius net' is the closure of {*A*, *B*, *C*} under the operation of harmonic conjugation.[22] Instead of defining the Möbius net merely as a closure of a certain set, the geometers of the time (Von Staudt, but also Klein)[23] distinguished various stages in the construction. Their goal was to define an isomorphism between the points of the net and the rational numbers. To give an idea of their reasoning, let me just explain how they coordinated integers to certain points of the real projective line. Let *A*, *B* and *C* be labelled 0, 1 and ∞ respectively. Then consider the harmonic conjugate to 0 with respect to 1 and ∞,

and label it '2'; then the conjugate to 1 with respect to 2 and ∞ will be called '3'; and the procedure can be repeated over and over without ever reaching the point ∞. Now, to construct the point −1, one just has to reverse the process: −1 is the conjugate to 1 with respect to 0 and ∞, −2 the conjugate to 0 with respect to −1 and ∞, etc. (see Figure 1.9).

In extending the operation in a suitable way, one can associate to each rational number a point of the Möbius net – and vice versa, to each point of the net, one and only one rational.[24] This coordination has of course many interesting properties. First, the ordinal relation between rational coordinates is compatible with projective separation, in the sense that two points of the net P and Q separates R and ∞ if and only if their respective coordinates p, q, and r are such that $p < r < q$. Second, one can associate with any four points P, Q, R, S of the net the quantity

$$\frac{r-p}{r-q} : \frac{s-p}{s-q}$$

This quantity is invariant by projective transformation, and it is equal to −1 when P, Q is a harmonic conjugate to R and S – in brief, this quantity has all the usual properties of cross-ratios. In fact, Von Staudt defined cross-ratio in this way, without relying on any prior concept of distance.[25]

Armed with this, Von Staudt attempted to derive the fundamental theorem. He noted first that a Möbius net is fixed as soon as three of its points are fixed. Now, he argued, as a projective transformation leaves the cross-ratio invariant, the image of every element of the net is fixed as soon as the images of the three original points are fixed. So far, so good – the trouble arose when Von Staudt misleadingly concluded from this that the image of every element of the line was fixed as soon as the images of

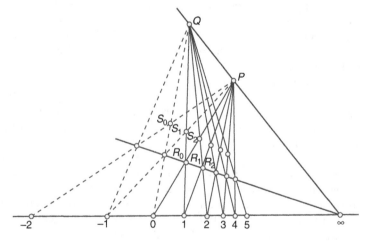

Figure 1.9   The introduction of (integral) coordinates on a projective line
Source: Klein (1928, p. 159).

the three original points (and thus of the points of the net they generated) were given. Of course, he recognized that they were points on the line that did not belong to the net. But he seemed to have believed that the density of the Möbius net was enough to guarantee that, as soon as the images of all the points of the net were defined, the images of the extra points would be fixed. In a modern (anachronistic) language, Von Staudt confused two distinct ordinal properties: density and density in a set. That a given ordered subset of the line is dense does not imply that it is dense in the line.[26] For deriving the fundamental theorem, this is the latter ordinal property which is needed. Indeed, if the net were dense in the line, one could view any extra point not in the net as a limit of a convergent series of elements of the net – one could thus associate any extra point with the real number that is the limit of the rational coordinates of the points belonging to the series converging to it. In Coxeter's axiomatization above, the role of the ordinal axioms, and especially of axiom 14, is precisely to enforce the fact that the Möbius net is dense in the line. But nothing, in Von Staudt, played the role of axiom 14. Klein (1873) was the first to uncover Von Staudt's mistake. The mathematician did not succeed right away in filling the gap, however, and a fascinating story began here, which would last until the opening of the twentieth century (see section 1.4).[27]

In section 1.1, I argued that the delimitation and scope of the 'new geometry' was not clear to the geometers of the first half of the nineteenth century. Certain properties related to incidence were invariant by projective transformations (at least when the Euclidean space was suitably extended); but ordinal and even metrical features (in the sense that they involved a reference to distances and angles) were also counted as 'projective'. It was thus difficult to characterize what the new theory was about. As I suggested above, this was not so much an issue as long as Euclidean geometry was regarded as the fundamental framework. But as soon as projective geometry was considered as an independent science (a possible basis for the other kinds of geometries, even), the question became central. Von Staudt made some advances. He succeeded in defining harmonic conjugation without resorting to distance or angle, and he even found a non-metrical characterization of rational cross-ratio. The method Von Staudt used was based on the quadrilateral construction and thus on the considerations of incidence relations in the real projective plane. His new definitions paved the way for a characterization of projective geometry in terms of incidence. Alas, his programme did not come to fruition. Klein showed that some key projective theorems (Pappus's and the fundamental theorems) could not be reached by Von Staudt's method. To understand the depth of the problem, let us recall that the fundamental theorem is needed to prove that the two rival notions of homography (as the product of perspectivities and as transformations which preserve cross-ratios) were coextensive. Von Staudt's mistake was to leave the very concept of projective transformation, and thus the whole topic of real projective geometry, indefinite. The situation was all the more disturbing as the theorems resisting Von Staudt's treatment did not seem to involve more than incidence. Pappus's theorem says something about how lines intersect in the real projective plane – no reference is made to order or to any metrical concept. Not only do incidence axioms not suffice to ground projective geometry, but some results about incidence seem to be grounded on some ordinal or continuity assumptions. All this cast doubt on the practicability of Von Staudt's programme. For Klein, projective geometry thus remained a hybrid theory, which combined Von Staudt's 'pure' method with some ordinal considerations. In Coxeter's axiomatic presentation, the heterogeneous character of the theory is reflected in the fact that two different groups of axioms are required to develop the whole body of the projective theory. So we see that the 'axiomatic turn' did not eliminate the problem concerning the nature of the projective invariants. On the contrary, it deepened it – the notion of group of axioms, and the rigorization of the proof, allowed geometers to understand that some key projective theorems were dependent on the ordinal assumptions. As we will see now, it is this issue concerning the proper definition of the subject matter of projective geometry which was behind Russell's PoM VI analysis.

1.3   Russell's twofold analysis

We come now to PoM VI. To oversimplify (without yet distorting the facts), one could say that, in chapter 45, Russell presents projective geometry as a pure theory of incidence relations, while, in chapter 46, he construes it as a pure theory of ordinal relations. Indeed, Pasch's approach (summarized in chapter 46) consisted of defining projective geometry in terms of betweenness and separation, while Pieri's account (related in chapter 45) amounted to deriving projective geometry from incidence alone. In other words, Russell, unlike Coxeter, did not regard projective geometry as a combination of two subtheories. He attempted to homogenize the conceptual framework and offered two reductions of the projective toolkits: to incidence relations in chapter 45 and to ordinal relations in chapter 46. The two axiomatic systems Russell developed did not then differ in some insignificant formal details – they displayed two completely different views of what projective geometry is. Russell's discussion did not concern the formal garment one should put on a theoretical body, but the body itself; it had to do with the specificity and the nature of geometrical thought. I will now explain the two approaches.

I will begin with the Pasch-inspired account that Russell labelled 'descriptive geometry'.[28] Coxeter maintained Russell's term and said (1947, p. 160):

> The ... *descriptive* geometry is of theoretical as well as practical interest, since the loss of the principle of duality is compensated by the fact that two points now determine a unique segment. Consequently the relations of

incidence and separation, instead of being undefined, are both expressible in terms of the single relation of 'three points' order, or *intermediacy*.

To show what is meant by defining incidence from the ordinal relation, I have listed the nine axioms and two definitions that Coxeter gives, omitting only the continuity axiom (1947, pp. 161–2):

1. There are at least three points.
2. If *A* and *B* are two points, there is at least one point *C* such that *B* lies between *A* and *C* (hereafter noted [*ABC*]).
3. If [*ABC*], then *A* and *C* are distinct.
4. If [*ABC*], then [*CBA*] but not [*BCA*]. **Definition:** The interval *AB* is the class of points *X* such that [*AXB*] with the points *A* and *B*; the ray *A/B* is the class of points *Y* such that [*BAY*]; the line *AB* consists of the interval [*AB*] with the rays *A/B* and *B/A*.
5. If *C* and *D* are two points on the line *AB*, then *A* is on the line *CD*.
6. There is at least one point not on the line *AB*.
7. If *A*, *B*, *C* are three non-collinear points, and *D* and *E* are such that [*BCD*] and [*CEA*], then there is a point *F* on the line *DE* with [*AFB*]. **Definition:** The plane *ABC* is the class of points collinear with the pairs of points on the intervals [*BC*], [*CA*], [*AB*].
8. There is at least one point not in the plane *ABC*.
9. Two planes which have one common point have another point in common.

This axiomatization is similar to Peano's (1889b), which is a formal version of Pasch's (1882).[29] It is roughly the same as those given by Russell.[30] As we can see, the geometry exposed here contains only two indefinables: the point and the relation between; all the other concepts (line, plane, incidence relations between lines, etc.) are defined in terms of these two. Coxeter is thus perfectly right to say that, in this framework, the incidence relation is derived from the relation of betweenness.

Of course, descriptive and projective spaces are not the same structure. Coxeter described the descriptive space as the restriction of the real projective space to a convex region, that is, the geometry of the descriptive space (plane) is the geometry which holds between the elements of a convex region of the real projective space (plane). For instance, an affine map of the projective plane is a descriptive plane.[31] Pasch, however, succeeded in showing how to 'complete' the descriptive space so as to obtain the projective space. Let me quote Russell's clear account of Pasch's construction (PoM, pp. 400–1):

An ideal point is defined as follows. Consider first the class of all lines passing through some point, called vertex. This class of lines is called a *sheaf* of lines ... A sheaf so defined has certain properties which can be stated without reference to the vertex ... All the properties of a sheaf which can be stated without reference to the vertex, are found to belong to certain classes of lines having no vertex, and such that no two of the class intersect. For these a simple construction can be given, as follows. Let *l*, *m* be any two lines in one plane, *A* any point not in this plane. Then the planes *Al*, *Am* have a line in common. The class of such lines for all possible points *A* outside the plane *lm*, has the properties above alluded to, and the word sheaf is extended to all classes of lines so defined. Thus, in Euclidean space, all the lines parallel to a given line form a sheaf which has no vertex. When our sheaf has no vertex, we define an ideal point by means of the sheaf.

In Figure 1.10, I have attempted to give a planar representation of the three-dimensional construction here described. The projective point *I* does not belong to the descriptive plane *Δ'*. Thus, the set of descriptive lines (such as *i*, *j*, *k*) that the projective geometer would characterize as being the pencil of lines passing through *I* cannot be so described. Pasch's achievement is to have succeeded in defining such a set of lines in a purely descriptive way, without referring to any non-descriptive (ideal) vertex. Having done this, Pasch identifies the projective points to these sets of descriptive lines. The procedure is similar to, if more complicated than, today's standard projective completion of the affine space.[32]

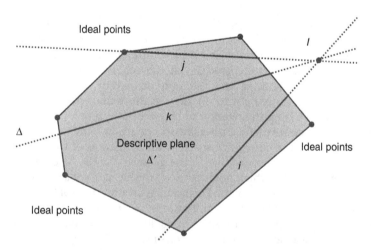

Figure 1.10   The descriptive plane *Δ'* is represented as a (shaded) convex part of the projective plane *Δ*. The descriptive lines correspond to segments of projective lines. All the points (like *I*) outside the descriptive plane are 'ideal points'. They can be represented as a certain set (containing *i*, *j*, *k*, etc.) of non-intersecting descriptive lines

I will not enlarge on this beautiful achievement, because Russell finally does not retain Pasch's option. What is important is not so much the detail of the construction as the fact that Russell was aware that it was possible to develop projective geometry from two indefinables, the point and the relation of betweenness.[33]

Seen from a distance, this approach could have suited Russell well. First, it allows him to characterize projective space without resorting to any reference to Euclidean and metrical notions. As I said at the beginning of this chapter, one of Russell's goals was to show that projective geometry was the most fundamental sort of geometry and that Pasch's construction, by making projective space independent of any metrical assumptions, gave it what he looked for. Second, betweenness, or order, is a key concept in Russell's thought. Part IV of PoM is entirely devoted to this type of relation, which is presented as being central to mathematics (pp. 199, 226). Pasch's reasoning could have been taken as a dazzling confirmation of this claim: order is not only the keystone of real analysis, but also of geometry. Surprising as it might appear at first, Russell let the opportunity slip and preferred to endorse the second definition of projective space – the one presented in chapter 45, based on the work of the Italian mathematician Mario Pieri (1898). What were the distinctive features of Pieri's reasoning?

From Coxeter's axioms, one can easily prove Darboux's lemma, according to which two points a, b do not separate two points c, d if, and only if, there are two points m, n such that both a, b and c, d are harmonic conjugates with respect to m and n. That a, b are harmonic conjugates with respect to m, n was noted in PoM '$aH_{mn}b$' – and Russell resumed the usual definition of the projective segment (acb) as the locus of point x such that a, b do not separate c, x (see Figure 1.7). With this notation, Darboux's lemma ran as follows:

$$D \in (acb) \Leftrightarrow \exists m \exists n \ (aH_{mn}b \ \& \ cH_{mn}d)$$

On the left-hand side of the equivalence, there occurs an ordinal concept (segment and thus separation); on the right-hand side, a reference is made to the harmonic conjugation, which could be explained away in terms of the quadrilateral construction, that is, in terms of incidence relations. This theorem seems therefore to throw a bridge between the two groups of axioms distinguished by Coxeter.

Pieri's great idea was to take advantage of this fact to elaborate a definition of the projective segment (the left-hand side of the equivalence) in terms of the harmonic conjugation (its right-hand side). More precisely, Pieri (1898, pp. 24–9) defined (acb) as the locus of the harmonic conjugate of c with respect to a variable pair of distinct points with respect to which a and b are also harmonic conjugates. As the quantificational structure of the definition is quite complicated (since it is of the $\forall\forall\forall\exists\exists$-type), I have given a translation of the equivalence in the Russellian notation:

$$\forall x \forall y \forall x' \forall y'\; (y' \in (xyx') \Leftrightarrow \exists z \exists z'\; (xH_{zz'}y\; \&\; x'H_{zz'}y'))$$

Thus, the fact that a point belongs to a projective segment (or the fact that two pairs of collinear points are separated) is reduced to the fact that there are certain points with respect to which some other points are conjugated – which is a pure incidence-theoretical fact, concerning how certain lines intersect in the plane. Russell explained the move quite clearly in a key passage of PoM (p. 385):[34]

> If four points x, y, x', y' be given, it may or may not happen that there exist two points a, b such that $xH_{ab}y$ and $x'H_{ab}y'$. The possibility of finding such points a, b constitutes a certain relation of x, y to x', y' ... Pieri has shown how, by means of certain axioms, this relation of four terms [harmonic conjugacy] may be used to divide the straight line into the two segments with respect to any two of its points, and to generate an order of all the points on a line.

In the standard presentation, resumed by Coxeter, real projective geometry arises from a combination of two different axiom groups, an incidence and an ordinal one. In Pieri's approach, real projective geometry is derived from only one indefinable relation, the incidence relation.

Things are a bit more complicated, however. As we have shown, Coxeter's axiom group of incidence does not suffice for deriving the whole body of projective geometry. In particular, the fundamental theorem cannot be derived from the incidence postulates. So, even if no primitive order relation (and no axioms of order) is introduced, something else must be given, if the fundamental theorem is to be proved. Pieri showed that three further axioms are needed to complete the construction. I give here the Russellian version of these postulates (PoM, p. 386):

1. If d is on the line ab, but does not belong to the segment (abc), and does not coincide with a or with c, then d must belong to the segment (bca).
2. If a, b, c be distinct points on a line, and d be a point belonging to both the segments (bca) and (cab), then d cannot belong to the segment (abc).
3. If a, b, c be distinct collinear points, and d a point, other than b, of the segment (abc), and e a point of the segment (adc), then e is a point of the segment (abc).

In spite of appearances, these postulates are not about segments or any other ordinal ideas, and their addition does not thus introduce any new indefinable. Indeed, using the definition of the segment in terms of incidence relations, the three axioms can be reformulated so as to be free of all references to ordinal notions. Yet, they allowed Pieri to use his derived ordinal notion of segment exactly as it is used in the 'standard' presentation, that is, as if order were a primitive notion. Pieri concluded his axiomatization with a continuity 'Dedekindian' axiom – here is the Russellian version (PoM, p. 387):

> If any segment (*abc*) be divided into two parts *h* and *k*, such that, with regard to the order *abc*, every point of h precedes every point of *k*, while *h* and *k* each contain at least one point, then there must be in (*abc*) at least one point *x* such that every point of (*abc*) which precedes *x* belongs to *h*, and every point of (*abc*) which follows *x* belongs to *k*.

From this and the other postulates (which, once again, involve only one indefinable concept, the incidence relation), the fundamental theorem of projective geometry and Pappus's follow. Russell was then right: even if Pieri introduced some new axioms, he nevertheless succeeded in reducing all the former indefinable relations to incidence alone. As Coxeter explained (1949, p. 33): '[Darboux's lemma] is especially significant since it would enable us to define separation in terms of incidence, instead of taking separation to be a second undefined relation ... This idea is due to Pieri, [who] reduced the undefined relations to incidence alone and reduced the axioms of order to the [three quoted above plus the Dedekindian postulate].'

I would like to pause awhile to consider Pieri's intellectual achievement.[35] It would seem at first that there is an insuperable intuitive and conceptual gap between ordinal and incidence notions. The way the points are located on a line seems indeed to have nothing to do with the way points and lines intersect on a plane. Pieri's beautiful result proves that this is not the case – it shows that, for ordering the points on a line, one is only required to set appropriate constraints on the way lines intersect in the plane. This idea is breathtaking: what, indeed, seems to be more ultimate than order? When Kant sought to prove that space was irreducibly phenomenal, he told us about the ordinal distinction between the right and the left. This shows that such an ordinal distinction was considered by him to be very basic and foreign to all conceptual frameworks. Even if, in Hilbert's *Grundlagen*, order was no longer viewed as pertaining to intuition, a similar opinion was expressed: ordinal concepts were separated from the incidence and the metrical ones in a particular 'group' – here as well ordinal notions occupied their own nest. Last, but not least, Part IV of PoM suggests that, for Russell too, ordinal relations were primitive and ultimate. It was precisely these ideas that Pieri attacked. What he challenged was the conceptual insularity of ordinal considerations – he showed that order could be regarded as a mere by-product, generated by the structure of the incidence relations. This disappearance of the ordinal postulates in Pieri's axiomatization, far from being a mere formal trick, has then a deep conceptual meaning.

But let me resume the story. Russell gave two accounts of projective geometry. In the first, it was regarded as a theory of points and betweenness. In the second, it appeared to be a theory of incidence. There is absolutely no doubt that Russell took the second account to be the most fundamental one. For him, Pieri, not Pasch, should be regarded as the true founder of projective geometry (PoM, p. 421):

> The true founder of non-quantitative Geometry is Von Staudt. It was he who introduced the definition of a harmonic range by means of the quadrilateral construction, and who rendered it possible, by repetitions of this construction, to give projective definitions of all rational anharmonic ratios ... But there remained one further step, before projective Geometry could be considered complete, and this step was taken by Pieri. In Klein's account, it remains doubtful whether all sets of four collinear points have an anharmonic ratio, and whether any meaning can be assigned to irrational anharmonic ratios. For this purpose, we require a method of generating order among *all* the points of a line ... This is effected by Pieri, with the help of certain new axioms, but without any new indefinables. Thus at last the long process by which projective Geometry has purified itself from every metrical taint is completed.

Russell's choice is surprising. As we have already remarked, Russell placed ordinal relations at the very heart of his thought, and this not only in his metaphysics, but also in his conception of mathematics. Now, we have just seen that Russell, from a technical point of view, had the means to construct projective geometry on a purely ordinal basis. Why did he refuse to do so then? Why did he appear in PoM VI to be willing to avoid order, while he seemed to have many reasons to advance it? Coxeter (1949, p. 33) remarked that Pieri's 'simplification is to some extent illusory, as these axioms would be quite complicated if we expressed them directly in terms of incidence'. He then asked: 'which is preferable: a number of simple axioms involving two undefined relations, or fewer but far more complicated axioms involving only one such relation?' His answer was that this is a matter of taste. Given the role played by ordinal notions elsewhere in PoM, how can we explain Russell's taste here?

1.4 Russell's choice

To answer this question, we need to turn to the genesis of Russell's thought. Describing in detail the evolution of Russell's thought from *An Essay* (1897a) to PoM is too big a task to be undertaken here, and I will content myself with giving an overview of a complicated gestation I have examined elsewhere in depth.[36] My point is that there is coherence in Russell's efforts: his 1903 emphasis on incidence is the direct result of his early involvement in Von Staudt's tradition of pure geometry.

In *An Essay*, Russell explained that the history of metageometry (i.e. non-Euclidean geometries) may be divided into three periods: the synthetic, the metrical and the projective. The first period corresponded to the works of Lobatchewsky and Bolyai, who explored axiomatically the consequences of the negation of the axiom of parallels. According to Russell, the 'metric stage', inaugurated by Riemann, had a much deeper import: '[it] regarded space as a particular case of the more general conception of a *manifold* [and] taking its stand on the methods of analytical metrical Geometry, it established two non-Euclidean systems, the first that of Lobatschewsky, the second ... a new variety, by analogy called spherical' (1897a, p. 18). The third and last period 'is chiefly distinguished from the second, in a mathematical point of view, by its method, which is projective instead of metrical' (ibid.). Russell's aim was to develop the philosophical interpretation of the third approach. As the quotes suggest, the contrast between the two last eras in the development of metageometry rested on the opposition between the direct synthetic approach inherited from the works of the pure geometers such as Poncelet, Chasles and Von Staudt, and the analytical Cartesian method that Russell saw resurfacing in Riemann.

As these disputes are nowadays forgotten, a few words about the opposition between analytic and synthetic geometry might be useful.[37] After the Cartesian reform, it was common to oppose the ancient geometry of the Greeks to the new one (which made use of coordinate systems and algebraic equations to represent geometrical facts). To resolve any given problem, the former used some piecemeal ad hoc tricks, where the latter proceeded according to a uniform and general path. What is more, some theorems, which could be easily demonstrated using coordinates and algebra, exceeded the power of the old synthetic approach. It is no surprise, then, if the new analytical method soon superseded the Greeks' way of doing geometry. At the beginning of the nineteenth century, however, several mathematicians sought to challenge the primacy of Cartesian geometry. Its success had left some questions unanswered: How could a reasoning developed for thinking about numerical values (algebra) produce such a surprising unity within geometry? According to the new geometers, the secret power of algebra needed a geometrical explanation, and one of the advantages of the ancient method was that it never lost sight of the geometrical aspect of the subject.[38] Projective geometry was erected in this highly polemical context: it was explicitly seen by Poncelet and his followers as an alternative to the Cartesian method, that is, as a theory which was as general as analytic geometry, while remaining faithful to the geometrical content of the subject. Von Staudt's is certainly the climax of this movement. As we have seen, he succeeded in introducing, using only geometrical resources (iteration of the quadrilateral constructions), numerical coordinates on a real projective line. The German geometer thus managed to deduce the possibility of doing Cartesian geometry from the geometrical properties of the projective space.[39]

In this debate, Russell clearly sided with the synthetic camp and maintained Von Staudt's approach.[40] This explained his dissatisfaction with Riemann, who based his investigation on the notion of numerical manifold. In *An Essay* (1897a), Russell opposed metrical geometry, described as a theory of quantity, with projective geometry, defined as a doctrine of quality, and sustained further that the concept of quantity presupposed the notion of quality. A pure non-quantitative characterization of projective geometry was thus needed, and Riemann's characterization of space did not fulfil this requirement.[41] How then did Russell define the notion of projective space in 1897? How did he connect the notion of quality to projective geometry? On this matter, *An Essay* is very disappointing.[42] Russell pretended to have identified the 'postulates' of the projective theory, but the 'axioms' he listed can hardly be taken as genuine ones, even when assessed according to the contemporary standard.[43] What is more, the link between these axioms and the notion of quality remained obscure. In an article published in the *Revue de Métaphysique et de Morale* (1899), Poincaré fiercely attacked Russell's reasoning. The French mathematician rightly pointed out that Russell's alleged axiomatization was mathematically empty. He also remarked that the (at the time) incipient topology was, more than the projective theory, entitled to be called a geometry of quality, since the qualitative invariance was built on the very definition of the topological transformations.[44] To summarize then, Russell (1897a) neither gave the axioms at the basis Von Staudt's reasoning, nor did he explain the link between projective transformations and the notion of quality.

A new approach is presented, two years later, in Russell's answer (1899a) to Poincaré's critical article. He gave up his notion of 'qualitative' geometry and delineated the nature of projective axioms more precisely. Here are the six postulates of his new system (1899b, pp. 403–4):[45]

Axiom I. There is a class *A* of objects ... such that any two of these objects, e. g. A_1, A_2, uniquely determine another object (a_{12} say) belonging to a different class *a* ... If the object a_{12} determined by A_1 and A_2 is not identical with the object a_{13} determined by A_1 and A_3, then the three objects A_1, A_2, A_3 uniquely determine an object (α_{123}) belonging to a new class α, which again does not determine uniquely the objects from which it is determined. Moreover a_{12} and α_{123} are independent of the order of the determining objects; and α_{123} is also determined by A_1 and a_{23}, or by A_2 and a_{31}, or by A_3 and a_{12}.

Axiom II. Two objects of class α (α_1 and α_2) determine uniquely an object $_{12}a$ of class a; and if $_{12}a$ is not identical with $_{13}a$, then α_1 and α_2 and α_3 determine uniquely an object $_{123}A$ of class A, which is also determined by α_1 and $_{23}a$. Two objects of class a, or four of class A or of class α, determine nothing. Thus all the objects determined by means of objects in the classes A, a, α belong in turn to these three classes.

Axiom III. When two objects α_{123}, α_{124} are respectively determined by A_1, A_2, A_3 and by A_1, A_2, A_4, the object of class a determined by α_{123} and α_{124} is the same as that determined by A_1 and A_2.

Axiom IV. Three objects α_{123}, α_{124}, α_{125}, determined respectively by A_1, A_2, A_3; A_1, A_2, A_4; A_1, A_2, A_5, collectively determine nothing. Three objects α_{123}, α_{145}, α_{167} (provided the first and the second do not determine the same object a as the first and third) collectively determine the object A_1.

Axiom V. Let a_{23} be the object determined by A_2 and A_3, and A the object determined by a_{23} and α_{145}. Then the object a determined by α_{123} and α_{145} is the same as that determined by A_1 and A_2.

Axiom VI. When two objects $_{123}A$, $_{124}A$ are respectively determined by α_1, α_2, α_3, and α_1, α_2, α_4, they determine together the same object $_{12}a$ as is determined by α_1 and α_2.

Russell also gave what he called an algebraic version of this system (ibid., pp. 404–5): instead of containing three classes of signs, it involved only one type of symbols and a product operation. For instance if 1, 2, 3 belonged to the class α_1, the products 12, 13, 23 belonged to the class a, and 123 to the class A. To grasp the idea behind Russell's reasoning, one must interpret the classes A, a, α as, respectively, the set of planes, the set of lines and the set of points (or, dually, as the sets of points, of lines and of planes). Now, when Russell says that 'A_1, A_2, uniquely determine another object (a_{12} say) belonging to a different class a', one should understand that two planes determine one and only one line (or, dually, that two points determine one line). In the algebraic version of the axiomatic presentation, this notion of determination corresponds to the product operation. The product then 'coded' the geometrical operations of 'intersection' and 'projection' between points, lines and planes (two points are said to be 'projected' in the line they determine). Viewed in this way, the six postulates appear then to put conditions on the incidence relations between the three kinds of items.[46] And Russell succeeded in showing that these conditions suffice to deduce Von Staudt's unicity theorem.[47]

Russell's system was in fact an elaboration from Grassmann's theory of progressive and regressive products. Russell learnt the doctrine through Whitehead's *A Treatise on Universal Algebra*, and the peculiarity of Russell's presentation can be traced back to this origin. The following quote from Whitehead (1898, p. 191) should suffice to show the relevance of the comparison ($v - 1$ is the dimension of the projective space, $\rho - 1$ and $\sigma - 1$ the dimensions of the projective subspaces P_ρ and P_σ that are multiplied):

If $\rho + \sigma < v$, then $P_\rho P_\sigma$ is progressive and represents the containing region ... of the two regions P_ρ and P_σ; unless P_ρ and P_σ overlap, and in this case the progressive product $P_\rho P_\sigma$ is zero.[48]

If $\rho + \sigma > v$, then $P_\rho\,P_\sigma$ is regressive and represents the complete region ... common both to P_ρ and P_σ; unless P_ρ and P_σ overlap in a region of order greater than $\rho + \sigma - v$, and in this case $P_\rho P_\sigma$ is zero.[49]

If $\rho + \sigma = v$, then $(P_\rho P_\sigma)$ is a mere number and can be considered either as progressive or regressive.

Whitehead's progressive product is thus the algebraic correlate of the geometrical operation of 'projecting', while the regressive product is the algebraic representation of the geometrical operation of 'intersecting'. For a reader of Whitehead's book, it was thus natural to associate the multiplication with the incidence relations between points, lines and planes. Note that Russell explicitly acknowledged Whitehead's influence in an important letter to Couturat dated 21 June 1900 (Schmid 2001, p. 182):

I liked your review of Whitehead's book very much, and Whitehead felt the same. However, I would like to point out what seems to me to be a mistake. Grassmann's algebra (even before the multiplication) is not one of completely pure projective geometry: intensity has an essential role in it; and, since it does not represent a mass, it could represent only a metrical idea. This is the reason why Whitehead, who agrees with me on this, speaks not of projective geometry, but of descriptive geometry. You will see that it proves V. Staudt's construction without the third dimension (p. 215) – it could not do that by a genuine projective method. The true algebra of projective geometry is the one I have devised for my answer to Poincaré: it contains only the multiplication. I have convinced Whitehead of this fact, and he will put this algebra in his second volume.

In Whitehead's theory, points, lines and planes were always associated with a scalar quantity (like the vectors of a vector space). Russell's move, as explained in the letter, consisted in extracting, from Whitehead's doctrine, the part corresponding to the progressive and regressive products. The aim of the system expounded in (1899a) was to disclose the Grassmannian products from its embedding in the more general Grassmannian algebra. For Russell, only this part of Whitehead's theory represented 'the true algebra of projective geometry', because only this part displayed the structure of the incidence relations between points, lines and planes in the real projective space. I have not the space here to fill in all the technical details[50] – what is important is to understand that Russell's focus on the progressive and regressive multiplication, and then on incidence relations, was an attempt to delineate what constituted the kernel of the projective method.

To summarize, in 1899, Russell, awoken by Poincaré, acknowledged the fact that the former characterization of projective geometry as a science of quality had absolutely no mathematical meaning. But the philosopher remained opposed to the 'quantitative' approach, according to which space was nothing other than a numerical manifold. The new idea was to base projective geometry on an axiomatization of incidence relations, which allowed him to deduce rigorously the unicity of Von Staudt's quadrilateral construction. In this second stage, accounting for projective space meant axiomatizing the incidence relations between projective points, projective lines and projective planes.[51]

However, a difficulty still burdened Russell's view. As we have said above, Klein had shown that Von Staudt's reasoning could not give what it claimed to give. Now, Russell writes that (1899a, p. 405) 'to show that [the above] axioms suffice, it is only necessary to prove ... the uniqueness of Von Staudt's quadrilateral construction, since all projective geometry ... follows from this construction'. Did Russell ignore the fact that certain crucial results (Pappus's and the fundamental theorem) remained out of Von Staudt's reach? No. In another passage, he adds (1899a, p. 409): 'there is only one proposition of importance which this method, so far as I know, is incapable of proving. This is the proposition that *all* points of a line can be obtained by this construction, and that there is no finite gap in the line'.

The mention of 'finite gaps' recalls Klein's criticism of Von Staudt's proofs. But, if Von Staudt's procedure is acknowledged as not being powerful enough, what are we supposed to do with the projective propositions that are not deducible by a reasoning à la Von Staudt? Russell remained vague on this crucial issue. Following Klein's wake, he seemed to believe that some ordinal considerations were required to complete Von Staudt's programme. But he was very hesitant about how to deal with order. Was order a purely projective concept, or was it a metrical notion? Russell did not answer this question. He seemed to consider (1899a, especially pp. 379–80, 'Projective Geometry is not essentially concerned with order or series') that order was a metrical concept, and that pure projective geometry should be restricted to only a part of what was usually regarded as belonging to projective geometry (the fundamental theorem and Pappus's would then belong to metrical geometry!). At other times, for instance in (1898d), Russell seemed to regard order as a mere projective notion, without explaining how to connect it with incidence relations. A central problem was thus left unresolved. How could one reconcile the idea that projective geometry should be separated from any metrical consideration, with the fact that Von Staudt's approach, based on the incidence relations, cannot provide us with an adequate foundation? It seems that something (order?) has to be brought in to complete the projective building. But how can we avoid blurring the purity of Von Staudt's reasoning when introducing these foreign elements?

Pieri's work allowed Russell to remove all the strains put on his analysis. As we have seen, Pieri managed to deduce the fundamental theorem (and thus the whole body of projective geometry), without introducing any new indefinables. With Pieri, said Russell (see the quotation on p. 36), 'the long process by which projective Geometry has purified itself from every metrical taint is completed' – and this achieved the aim of Von Staudt's programme.[52] In PoM VI, Russell still placed Von Staudt's quadrilateral construction (i.e. the incidence relations) at the centre of the projective world. But, henceforth, he no longer had any problem with Klein's criticism. Thanks to Pieri, the severe technical difficulty which burdened his 1899 approach no longer existed. Taking into account the development of Russell's view about geometry allows us to explain why he chose Pieri's rather than Pasch's way of defining projective space. Betweenness is a concept that Von Staudt's followers would certainly have found suspect. Order is not a metrical notion – but at the same time, it cannot be easily related to Von Staudt's method, grounded on incidence between points, lines and planes. Thus, even if order played a central role in his thought, Russell's early commitment to Von Staudt's approach explained his decision to follow Pieri rather than Pasch: Pieri's work gave Russell the technical means called for by his conception of space and geometry.

1.5   Space as an incidence structure

The reader might have the impression that we have dwelt on the detail of Russell's work at the complete expense of any of its philosophical significance. Russell showed favour to Pieri's rather than to Pasch's account of non-quantitative geometry – so what? In this section, I will explain that this point is absolutely crucial for grasping Russell's general view of space, as it is developed in chapter 54 of PoM VI, and maintained by Whitehead in his neglected *The Axioms of Projective Geometry* (1906b).

Many commentators credit Russell (PoM) with an 'if-thenist' view of geometry. I will expound this interpretation in the two next chapters, but one part of the thesis, which I want to deal with straightaway, is that Russell would have adhered to a formal conception of the notion of space, according to which it is nothing else than a model of an axiomatic system. Received opinion has it that Hilbert (1899) was responsible for a radical break in mathematics: instead of regarding geometry as a scientific description of pre-given object (space), he considered that Euclidean space was defined by the postulates of the axiomatic system. For him, geometry was not the description of something (space); a geometrical theory was only a conceptual scaffolding which could be fleshed out in numerous different ways. Since I want to stay focused on Russell, I do not want to enter into the issue as to whether or not this picture was really endorsed by Hilbert.[53] According to the 'if-thenists', Russell would have accepted this view. Indeed, they claim that his logicist conception of geometry amounts to no more than saying that if T is the set of axioms of a geometrical theory, and p is a geometrical theorem, then $T \Rightarrow p$ is a logical truth. 'If-thenism' is just a way of expounding the axiomatic turn – of saying that geometrical space is completely characterized by the axioms T. As I will expound it, I do not think that Russell was an 'if-thenist'. Russell worked within an axiomatic framework; but he did not content himself with the idea that space is a conceptual scaffolding; he wanted more; he wanted to identify which conceptual scaffolding space is. He did not deny that space is a structure (or as he called it a 'relational concept') which satisfied a certain axiomatic system. But his problem was how to disclose which relational concept it is. PoM VI is entitled 'Space', and it aims at defining what the real subject matter of the geometrical sciences is. This demand is foreign to the Hilbertians (according to received opinion) and to the if-thenists, who rest content with any axiomatization of geometry. As a follower of Von Staudt that Russell was, there was absolutely no doubt that geometry constituted an independent branch of knowledge endowed with its own method, its own problems and its own concepts. The issue was then to bring out what the essence of geometry is, or, in other words, what the common core is of all the various axiomatic geometrical systems. This is precisely the goal Russell pursued in chapter 54. Let me quote one of its key passages (PoM, p. 372):

> As a branch of pure mathematics, Geometry is strictly deductive, indifferent to the choice of its premises and to the question whether there exists (in the strict sense) such entities as its premises define. Many different and even inconsistent sets of premises lead to propositions which would be called geometrical, but all such sets have a common element. This element is wholly summed up by the statement that geometry deals with series with more than one dimension. The question what may be the actual terms of such series is indifferent to Geometry, which examines only the consequences of the relations which it postulates among the terms. These relations are always such as to generate a series of more than one dimension, but have, so far as I can see, no other general point of agreement ... At present, I shall set up, by anticipation, the following definition: *Geometry is the study of series of two or more dimensions.*

As a branch of pure mathematics, 'geometry is strictly deductive', and so can always be expressed by an axiomatic system. Russell also recognizes that 'many different and even inconsistent sets of premises lead to propositions which would be called geometrical'. Does he, for all that, hold an 'if-thenist' position? No, because the important issue for him was not to say that the implication $T \Rightarrow p$ is a logical truth – it was to define what the different sets

T of geometrical premises have in common. Now, how did Russell answer this last question? How did he define the common core of all geometrical axiomatic systems?

In the passage above, Russell claimed that space is a 'series of two or more dimensions'. A few pages later, he explained that an n-dimensional series can be inductively defined from the $(n-1)$-dimensional series in the following way: an n-dimensional series is a series whose elements are an $(n-1)$-dimensional series (PoM, p. 374). In other words, he was relying heavily on serial order to elaborate his inductive definition of an n-dimensional series. And this renders his definition quite awkward, since the paradigmatic example of multiple series he is going to take is the projective space, where the lines precisely cannot be regarded as open series (the projective lines are topologically closed; see Figure 1.7). I think, however, that, when replaced in its context, Russell's idea is pretty clear. In the first page of PoM VI, he wrote that, in the previous parts of the book, the 'philosophical theory of one-dimensional series' has been completed, but that 'large branches of mathematics have remained unmentioned', such as geometry and complex number theory. Space and complex numbers appear then to have an additional structure, not reducible to that of series. Seen from this standpoint, what appears important in the definition of space (and of the complex number field)[54] as 'multiple series' is thus not the concept of 'series' (whose analysis has already been completed in part IV and V), but rather the idea of multiple dimensions. Now, what distinguishes multiple series from one-dimensional series is the fact that, in the former case, certain elements are common to two different series. That is, multiple series display incidence relations in the way one-dimensional series do not. In other words, what Russell was (a bit awkwardly) trying to say with his notion of multiple series is that the common point of all geometrical systems is to contain incidence axioms – and that space is then, at bottom, an incidence structure. According to him, to have a space, it would not be required to have metrical relations between points; it would not even be required to have order relations – what would be needed, on the other hand, is to have incidence relations between different spatial elements.

Of course, this reading fits well with the analysis of Russell's conception of projective geometry which we have made above. But to avoid being accused of tacking my interpretative grid onto chapter 44, let me substantiate the claim with an independent source. According to Whitehead, his *The Axioms of Projective Geometry* (1906b) is nothing else than an extension of Russell's PoM VI theory.[55] Here is the definition Whitehead gave of geometry at the beginning of his book (1906b, pp. 4–5):

Geometry, in the widest sense in which it is used by modern mathematicians, is a department of what in a certain sense may be called the general science of classification. This general science may be defined thus: given any class of entities *K*, the subclasses of *K* form a new class of classes, the science of classification is the study of sets of classes selected from this new class so as to possess certain assigned properties. For example, in the traditional Aristotelian branch of classification by species and genera, the selected set from the class of subclasses of *K* are (1) to be mutually exclusive, and (2) to exhaust *K*; the subclasses of this set are the genera of *K*; then each genus is to be classified according to the above rule, the genera of the various genera of *K* being called the various species of *K*; and so on for subspecies, etc. The importance of this process of classification is obvious, and is sufficiently emphasized by writers on Logic ...

Geometry is the science of cross-classification. The fundamental class *K*, is the class of points; the selected set of subclass of *K* is the class of (straight) lines. This set of subclasses is to be such that any two points lie on one and only one line, and that any line possesses at least three points. These properties of straight lines represent the properties which are common to all branches of the science which usage terms Geometrical, when the modern Geometries with finite numbers of points are taken account of ...

In Projective Geometry the subject viewed simply as a study of classification has great interest. Thus in the foundations of the subject this conception is emphasized, while the introduction of 'order' is deferred.

Whitehead first introduces the set-theoretical concept of 'classification'. This is a family of subsets defined on a given set. He then defines an Aristotelian classification of a set *X* as a partition of *X* (it is 'Aristotelian' because this sort of classification is the abstract schema of the division in species and genera). There is however another kind of classification that Whitehead labels 'geometrical' or 'cross-classification'. A cross-classification of a set *X* is such that (1) any two elements of *X* determine one and only one subset (line) of the classification, and (2) any such subset contains at least three elements of *X*. In other words, unlike what is the case in the Aristotelian classification, an element *x* of *X* can belong to different subsets (lines) of a geometrical classification – that is why this sort of family is labelled 'cross-classification'. See Figure 1.11 for an example of a cross-classification defined on a finite set.

Now, Whitehead defined geometry as the study of cross-classification. This characterization has of course much in common with Russell's definition of space as 'multiple series'. But it is more satisfying than Russell's, since it does not make any reference to order and links from the very beginning the concept of space with the axioms of incidence (Whitehead uses the term 'cross-classification axioms' to designate what we usually call 'incidence axioms'). Whitehead's definition is furthermore perfectly coherent with Pieri's approach, resumed in the rest of the book. As he noted at the end of the quote, 'in the foundations of the [projective geometry] this conception [of geometry as a science of cross-classification] is emphasized, while the

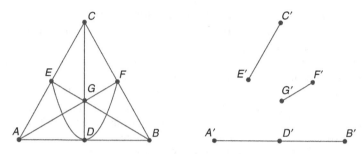

Figure 1.11 To the left is a cross-classification of the set X of seven points (this is the projective plane of order 2). One can check that the two conditions given by Whitehead are fulfilled by the seven lines, each composed of three points, of the classification. To the right is an Aristotelian classification of the same set X in three subsets

introduction of "order" is deferred'. That Whitehead's Russell-inspired tract explicitly connects the idea of 'multiple series' and 'cross-classification' with Pieri's development thus substantiates my interpretation of PoM 44.

Before closing this chapter, let me insist on the mathematical interest of this incidence-based approach. Today, incidence structure is a central concept in combinatorial geometry.[56] An incidence structure can be defined as a triple $C = \langle P, L, I \rangle$, where P is a set of 'points', L is a set of 'lines' and $I \subseteq P \times L$ is the incidence relation. If $(p, l) \in I$, the point p is said to lie on line l. Further constraints can be added to define more interesting structures. One of the most basic notions is the one of a (v, k, λ)-balanced incomplete block design (abbreviated hereafter to (v, k, λ)-BIBD). Let v, k, λ be three positive integers, such that $v > k \geq 2$; a (v, k, λ)-BIBD is an incidence structure $\langle P, L, I \rangle$, such that (1) the cardinality of P is v, (2) each line ('block') contains exactly k points, (3) every pair of distinct points is contained in exactly λ blocks. The definition does not perfectly match Whitehead's cross-classification, but, when $\lambda = 1$, the third condition corresponds to his first constraint (that any two points determine one and only one line).[57] I do not want to argue that Russell's and Whitehead's definition of space anticipates the development of the combinatorial theory of incidence structure and block design. What this comparison suggests, however, is that their thought was conceptually linked to the development of the modern combinatorial point of view which also emerged at the beginning of the twentieth century. In the passage quoted above, Whitehead alluded to the 'modern geometries with finite numbers of points', and he later referred (1906b, p. 23) to Veblen and Bussey's paper entitled 'Finite Projective Geometries', published in 1906. This work was in fact the culmination of a line of research whose main feature was to merge combinatorial considerations with geometrical developments. In 1896, E. H. Moore

(Veblen's teacher) published a long paper (referred to in Veblen and Bussey 1906) in which he defined for the first time the general and abstract notion of an incidence structure (which he called a 'configuration').[58] I am not claiming that Whitehead or Russell knew these works well – their main reference remained Pieri. But these developments were witness to the fact that the connection between space and incidence ran deep in the mathematical and philosophical world at the turning of the twentieth century.

1.6 Conclusion

In the first section of PoM VI, Russell presented two distinct axiomatizations of real projective geometry: the indefinables of the Pieri-inspired geometry (developed in chapter 45) are points, lines, planes and the incidence relations; the indefinables of the Pasch-inspired geometry (expounded in chapter 46) are points and the ordinal relations between them. The two characterizations[59] define exactly the same geometrical structure: the two systems allow us to derive exactly the same propositions – Desargues's, Pappus's and the fundamental theorem of projective geometry. So Russell was presenting two distinct 'formally adequate' analyses of projective geometry. Had he adhered to the view that the only criteria which have to be taken into account in the assessment of an analysis are the formal ones, he would have regarded the two characterizations as equivalent from a philosophical point of view. But he did not do so. In PoM VI, Russell made clear that he considered Pieri's way of construing projective space was far better than Pasch's. Pieri's construction, which based projective space on incidence, was rooted in a tradition that aimed at uncovering what was specific to the geometrical sciences. Russell saw himself as a member of his movement: his task was not merely to find *a* formal characterization of (projective) space, but to delineate *the* real essence of geometrical thought. And seen from this perspective, the definition of space as an incidence structure appeared to him as a better option than Pasch's ordinal characterization.[60] I will come back to the reasons behind Russell's choice. But for now, what is important is to realize that he was not just interested in formalizing geometry – the way this task was done mattered at least as much as its bare possibility.

This can be illustrated in another way. To define projective space, some mathematicians of the time (Klein, for instance) used homogeneous coordinates, which assimilate the real projective space to (what we today call) the quotient $\mathbb{R}^4/\sim$, where $(x, y, z, w) \sim (x', y', z', w')$ if and only if there is a real number λ, such that $(x, y, z, w) = (\lambda x', \lambda y', \lambda z', \lambda w')$.[61] In brief, some mathematicians did projective geometry analytically, in a refined Cartesian way (for Russell, they belonged to the 'Riemannian era'). By the end of PoM V, Russell had defined the structure $\mathbb{R}$ in purely logical terms. Why did he not then proceed from there by characterizing projective space as a numerical manifold? Why did he not espouse the analytic viewpoint? This was certainly the easiest way of reducing projective geometry to logic. Were Russell only interested in finding a formally adequate analysis of geometry, this would certainly have represented the best deal. Now, in PoM VI, this possibility is not even evoked. And what we have said explains why it is not: Russell sought to delineate what made geometry specific, and for a mathematician of the end of the nineteenth century this amounted to showing what was missing in the 'impure' Cartesian approach to geometry. The dismissal of the analytic construction of projective space, as the detailed discussion of Pieri's versus Pasch's characterization of projective geometry, gives substance to what I said in the Introduction: Russell, in PoM, did not content himself with a formal adequate analysis of geometry; rather, he asked for more, for a logical definition which accounted for the specificity of geometrical thinking.

2
Metrical Geometry

In Chapter 1, I said several times that 'space' and 'projective space' were for Russell synonymous. Yet, projective geometry is only a part of geometry – metrical features, whether Euclidean or non-Euclidean, do not belong to the projective setting. How did Russell deal with metrical notions? What view did he hold about metrical space? The emergence of non-Euclidean geometries created a considerable crisis within the philosophical world of the second half of the nineteenth century.[1] Kant, whose epistemological work still occupied a central place, founded his theory of the transcendental subject on the apodicticity of Euclidean geometry. The idea that non-Euclidean spaces were non-contradictory appeared then, if not as a direct refutation, at least as a challenge to Kantian views. And until late in the nineteenth century, some philosophers made a tremendous effort to restore the conceptual superiority of Euclidean geometry. From his *An Essay on the Foundations of Geometry* (1897a) onward, Russell accepted the existence of non-Euclidean geometries and fiercely criticized his fellow philosophers for not acknowledging the legitimacy of 'metageometry' (his name for non-Euclidean geometries). Scholars are prone to insist on this point. They usually do not, however, stress the relation between his acceptance of non-Euclidean spaces and his belief that projective space is the common core of the various metrical spaces. In this chapter, I will argue that one cannot understand the conception of metrical geometry expounded in PoM VI without resorting to Russell's critical reception of Klein's projective derivation of metrical concepts.

In the opening to Chapter 1, I said that PoM VI could be divided into three distinct parts: chapters 44–46 are devoted to projective space; chapters 47–49 to metrical geometry; and chapters 50–52 developed some philosophical considerations on space. Here, I will mostly focus on the second part. In chapters 47 and 48, Russell presented three different conceptions of metrical space; chapter 49 is more general in character, aiming to show that various kinds of space (not all metrical) are definable in purely logical terms.

So I will concentrate my analysis on chapters 47–48 which are explicitly devoted to the theory of metric. This point is important since most scholars take the opposite course and highlight certain passages in chapter 49 to the detriment of the mainline developments expounded in chapters 47–48.

In section 2.1, focusing on an important paper of Musgrave (1977), I will present what could be called the 'standard' interpretation of PoM VI, according to which Russell's view boiled down to an axiomatic 'if-thenist' approach to geometry. The rest of the chapter can be seen as an attempt to refute this reading. In sections 2.2 to 2.5, I will present the three different conceptions of metrical geometry Russell developed in chapters 47 and 48 of PoM. The first one (dealt with in section 2.2) was Klein's projective theory (1871). Klein's idea was to define the three classical distances (Euclidean, elliptic and hyperbolic) as a certain function of the cross-ratio. This view presented the advantage of supplying a common projective framework to what was at the time regarded as the three main metrical geometries. But in Russell's eyes it had the inconvenience of construing distance as a very complicated notion. The two other ways of defining metrical concepts are presented in section 2.3. First, Russell mentions the formal possibility of providing a direct axiomatic of distance. Second, he expounds the view that distance could be regarded as an empirical concept and, accordingly, that metrical geometry can be seen as an empirical science. In section 2.4, I will attempt to deepen our understanding of this last approach by comparing it to Poincaré's conventionalist view. In section 2.5, I will explain why Russell took the decision to favour the empirical conception of metrical geometry. This choice is really astonishing: while he had the means to integrate metrical geometry in the sphere of the logical sciences, he preferred, despite his logicism, to consider it as an empirical science.

The survey which follows is linked to the developments I describe in Chapter 4, which is devoted to the 1903 theory of quantity. Indeed, for Russell, metrical concepts are paradigmatic cases of quantitative notions (it is so much so that metrical geometry is often called quantitative geometry); accordingly, the two accounts of metrical geometry and quantity go hand in hand in PoM. In section 2.3 below, I will explicitly refer to Chapter 4 – but more generally the content of this chapter should not be kept separate from what I say in Chapter 4.

2.1   Logicism, if-thenism and non-Euclidean geometries

In his seminal paper (1977), Alan Musgrave distinguished two kinds of logicism: a strong form, according to which logicists aim at proving all the classical mathematical theorems from logical axioms, and a weak version, according to which logicists aim at proving the conditional propositions whose antecedents are the axioms of a mathematical theory and whose consequents are the various theorems that follow from them. Musgrave's interesting thesis[2] is that, since the strong form turns out to be a failure (because of the logical paradoxes), Russell, and the logical positivists after him, adopted the weak position, without yet renouncing the rhetoric of the strong version.[3] I will come back to this general interpretation in the next chapter. In this section, I will only focus on Musgrave's analysis of Russell's theory of geometry in PoM VI. Musgrave (1977, pp. 109–10) writes:

> It was actually Russell who found a way to rescue logicism from defeat; and the key to it was provided by the problem of assimilating *geometry* to logic. Frege had actually excluded geometry from the logicist thesis, and had endorsed Kant's view of it … Russell, on the other hand, thought that the discovery of non-Euclidian geometries had undermined Kant's original position, and in his first major publication he tried to rescue it. In his *Foundations of Geometry* of 1897, Russell sought what was common to Euclidean and non-Euclidean systems, found it in the axioms of projective geometry, and took a Kantian view of them. As for the additional axioms which distinguished Euclidean from non-Euclidean systems, these were empirical statements (Russell (1897), Introduction, section 9). But after he had adopted the logicist thesis, Russell sought a way to bring geometry into the sphere of logic. And he found it in what I shall call the *If-thenist manoeuvre*: the *axioms* of the various geometries do not follow from logical axioms (how *could* they, for they are mutually inconsistent?), nor do geometrical *theorems*; but the *conditional statements linking axioms to theorems* do follow from logical axioms. Hence geometry, *viewed as a body of conditional statements*, is derivable from logic after all.

Musgrave refers here to this passage of the introduction to the second edition of PoM (1937, p. vii):

> It was clear that Euclidean systems alike must be included in pure mathematics, and must not be regarded as mutually inconsistent; we must, therefore, only assert that the axioms imply the propositions, not that the axioms are true and therefore that the propositions are true.

But one also finds in PoM VI (in chapter 49) some passages that go in that direction. Musgrave continues (1977, p. 110):

> Russell argued that the discovery of non-Euclidean geometries forced us to distinguish *pure geometry*, a branch of pure mathematics whose assertions are all conditional, from *applied geometry*, a branch of empirical science. After describing the emergence of non-Euclidian geometry, he says (1903, p. 373): 'Geometry has become … a branch of pure mathematics, that is to say, a subject in which the assertions are that such and such consequences follow from such and such premises, not that

entities such as the premisses describe actually exist.' In this way the axioms of the various geometries cease to be problematic for the logicist, because they cease to be asserted as axioms at all (let alone asserted to be derivable from logical axioms): 'The so-called axioms of geometry, for example, when Geometry is considered a branch of pure mathematics, are merely the protasis in the hypotheticals which constitute the science. They would be primitive propositions if, as in applied mathematics, they were themselves asserted; but so long as we only assert hypotheticals ... in which the supposed axioms appear as protasis, there is no reason to assert the protasis, nor, consequently, to admit genuine axioms'. (Russell 1903, p. 430)

What emerges from this is the idea that Russell, in PoM VI, broke with his earlier views expounded in (1897a) and no longer shared the Kantian conception. That Russell, in PoM, rejected any reference to a non-logical a priori knowledge is a fairly uncontroversial claim. But Musgrave goes further than that: he claimed that Russell changed his view about the relation between metrical and projective geometry, and also that he no longer adhered to the idea that metrical geometries are empirical. Let me explain these two points in turn.

First, while Russell attempted in 1897 to bring out what was common to the various metrical theories, his starting point in 1903 was that the various metrical geometries were not compatible. The new issue for him was the following: how can we sustain geometry as a branch of logic, if the axioms of the various theories are inconsistent? According to Musgrave, Russell's answer to this is if-thenism: although the various geometrical axioms do not follow from logic, 'the conditional statements linking axioms to theorems do follow from logical axioms'.

Second, while Russell believed in 1897 that metrical geometries were empirical, in 1903 he sought 'a way to bring geometry into the sphere of logic'. (By 'geometry', Musgrave makes it clear that he means 'metrical geometry'.) More precisely, Russell, says Musgrave, would introduce in PoM the distinction between pure and applied geometry as a by-product of if-thenism: pure geometry is a body of conditional statements which were applied as soon as the antecedents (the axioms) are regarded as empirical truths. However, Russell in PoM VI would defend a logicist view of metrical geometry.

Among scholars, this reading has been dominant. In particular, the idea that the if-thenism is a necessary consequence of the multiplicity of possible metrics has been widely accepted.[4] Scholars usually base their interpretation on the passages Musgrave referred to (that is, those contained in the preface of the second edition, and those occurring in chapter 49 of PoM VI). Here, I will show that by taking into account the content of chapters 47 and 48, which bear more directly on the issue at stake, we will be led to challenge the basic tenets of Musgrave's picture and to oppose the two points above. Concerning the second point, I will claim that Russell in PoM did not see metrical geometry as a logical science, but as a piece of empirical knowledge. It is true that he developed the means to define distance in logical terms, and thus to consider metrical geometry as a part of pure mathematics. But for reasons that I will explain, he thought that this was not a good way to conceive metrical notions. Concerning the first point, I will show that Russell, when he considered metrical geometry as a pure logical construction, brought forward the organic unity that the projective viewpoint conferred to the whole subject. Far from emphasizing the incompatibilities between the various axiomatics, Russell insisted on the common projective root of the different theories. In a more general way, I am not convinced at all by Musgrave's idea that there is a break between 1903 and 1897 – I see Russell's conception of metrical geometry in PoM as a refinement and an extension of the views presented in *The Essay*.

Before entering into the detail of the discussion, let me add a few words on the internal difficulties posed by Musgrave's reading. Coffa pushed Musgrave's analysis to its breaking point (1981, p. 252):

> If the claim that mathematics is logic is basically no more than the claim that each conditional that has an appropriate set of mathematical axioms as antecedent and one of its theorems as consequent is provable in logic, then we can establish the reducibility to logic not only of mathematics but of a large number of obviously non-logical disciplines as well. Russell seems to be happy to acknowledge that 'pure physics' … is a 'branch of pure mathematics' … and hence of logic. But also pure biology, pure economics, pure geography, and, indeed, any axiomatizable (first-order) theory is, in Russell's odd sense, reducible to logic. If this is all that logicism claims then it would appear, both its significance and its relevance to any traditional philosophical issues … seem to be very much in doubt.

For Coffa, Russell's conditional logicism falls prey to a trivialization argument. If if-thenism is all that logicism claims, then an axiomatic geography, if such a thing exists, would become a logical science. This would deprive Russell's view of any philosophical significance. In other words, for Coffa, Russell, in order to accommodate the plurality and incompatibility of metrical geometries in the logical framework, had gone too far in weakening the logicist project. Now, Coffa is drawing correct conclusions from Musgrave's interpretation: if one wants to save Russell's work from such an immediate collapse, one then has to challenge Musgrave's reading. I mention the trivialization argument here only to point out that the detailed analyses which follow have a real importance for my interpretation of Russell's project as a whole (I will come back to the trivialization challenge in Chapter 3).

2.2   The projective definition of a metric

Russell's presentations in PoM chapters 47 and 48 are intricate and difficult to read. One thing is clear though: three ways of defining a metrical space are successively considered. Russell speaks first about the attempt at constructing a metric space from the sole notion of distance (from § 392 to § 395 of chapter 47), before passing on to the theory which considers metric as an empirical concept (from § 396 to § 399 of chapter 47, and some parts of chapter 48),[5] and finally presents the projective theory of distance (chapter 48). But if the degree of importance is considered, the order must be reversed: the most fundamental theories of metric space are the two last ones. I will begin by presenting the projective view of metrical space and postpone the presentation of metrical geometry as an empirical science and the axiomatic theory to the next section.

The projective theory of metric is developed by Klein in his important 'Ueber die sogenannte Nicht-Euklidische Geometrie' (1871). He extended in an unexpected way the previous works of Cayley (1859). Let me quote the introduction (1871, pp. 69–70):

> Cayley is concerned with showing that ordinary (Euclidean) geometry may be regarded as a special part of projective geometry. For this purpose, he sets up the general projective measure and then shows that its formulae become the formulae of ordinary geometry when the fundamental surface degenerates to a particular conic section, the imaginary circle at infinity. Our goal, on the other hand, is to explain the geometric content of the general Cayley measure as clearly as possible, and to recognize, not only that Euclidean geometry results from its specialization, but more importantly that it is related in exactly the same way to the various metric geometries arising from the different theories of parallels.

In other words, as Russell explains, 'the so-called projective theory of distance aims at proving that metrical (Euclidean and non-Euclidean) is merely a branch of projective Geometry' (PoM, p. 422). Since Russell, in chapter 48, closely follows Klein's reasoning, I will briefly explain its main outline.

Klein (1871, § 2) gives three formal properties of oriented distances on a line:[6] a distance is additive, that is, A, B, C being any three points on the line, dist(AB) + dist(BC) = dist(AC); the distance between two coincident points should be 0; dist(AB) should be the opposite to dist(BA). In section 3, Klein shows that, when two points a and b are fixed, and when the cross-ratio between two points x, y on the line (ab) is positive, then the logarithm of the cross-ratio (xy/ab) has all the required formal properties. As Russell explains (PoM, p. 422):

> Let us consider the anharmonic ratios of all ranges $axby$, where a, b are fixed points and x, y variable points on our line. Let α, ξ, β, η be the

coordinates of these points. Then $\frac{\xi-a}{\xi-\beta}:\frac{\eta-a}{\eta-\beta}$ will be the anharmonic ratio of the four points, which, since a, β are constants, may be conveniently denoted by $(\xi\eta)$. If now ζ be the coordinate of any other point z, we have $(\xi\eta)(\eta\zeta)=(\xi\zeta)$. Hence, $\log(\xi\eta)+\log(\eta\zeta)=\log(\xi\zeta)$. Thus the logarithm of the anharmonic ratio in question has one of the essential properties of distance, namely additiveness ... We have also $\log(\xi\xi)=0$ and $\log(\xi\eta)=-\log(\eta\xi)$, which are two further properties of distance. From these properties ... it is easy to show that all properties of distances which have no reference to the fixed points a, b belong to the logarithm in question. Hence, if the distances of points from a and b can also be made ... to agree with those derived from the logarithm, we shall be able to identify distance with this logarithm. In this way – so it is contended – metrical Geometry may be wholly brought under the projective sway; for a similar theory applies to angles between lines or planes.

The idea is thus to define the distance between any two points x, y on a given line as a function (a logarithm) of the cross-ratio between four points, x and y and two fixed points, that Klein calls the fundamental points.

So far, Klein has simply maintained the path taken by Cayley in his elaboration of a general projective metric. But he went further when he understood that the nature of the metric introduced on the line depended on the choice of the couple of fundamental points. One thing should be noted before presenting this development: Klein (like Cayley) was an analytic geometer who worked with homogeneous coordinates taking values in the set of complex numbers. This is important because he stated that the fundamental points must be the roots of a quadratic form $\Omega=ax_1{}^2+bx_2{}^2+2cx_1x_2$, where (x_1, x_2) is the (possibly complex) homogeneous coordinate of a point on the line, and a, b, c are real numbers. On the complex line, there are two points (not necessarily distinct) which make the form Ω equal to 0. Three cases can occur: the hyperbolic case, in which two distinct real points A and B are the roots of $\Omega=0$; the parabolic case, in which there is only one double root C; and the elliptic case, in which $\Omega=0$ has two imaginary conjugate complex points D and E. Figure 2.1 summarizes the situation.

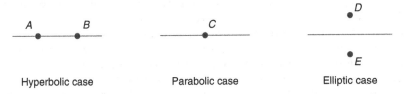

Figure 2.1 The three possible positions of the fundamental points in relation to the real part of the complex line
Source: Klein (1928, p. 81).

Now, Klein showed that (i) if the fundamental points are real and distinct, the projective metric defined on the real part of the complex line is a hyperbolic distance; and that (ii) if the fundamental points are the complex conjugates, the metric is elliptic. The intermediate case is more difficult, since, X and Y being any two real points distinct from C, the cross-ratio (XY/CC) is identical to 1, and therefore its logarithm is always equal to 0. But, as Russell explained, 'by a familiar process of proceeding to the limit' (PoM, p. 424), one can show that the difference between the coordinates of X and Y has all the formal properties of a distance. Klein then considered that the parabolic case corresponded to the Euclidean metric. After having dealt with the line (1871, §§ 2–7), Klein extended his construction to the plane (§§ 8–15), and to three-dimensional space (§ 16). In the plane, the fundamental form Ω has the following analytical expression:

$$\Omega = ax_1{}^2 + bx_2{}^2 + cx_3{}^2 + 2dx_1x_2 + 2ex_1x_3 + 2fx_2x_3$$

where a, b, c, d, e, f are real parameters. The *locus* of the points (x_1, x_2, x_3) that makes Ω equal to 0 is a (complex) conic, called the Absolute. We find here again the same division into three cases: if the Absolute is a real conic, then the metric is hyperbolic; if the Absolute is an imaginary conic, then the metric is elliptic; and if the Absolute is a degenerate conic which takes the form of a couple of complex points (the so-called circular points at infinity), then a new notion of a distance can be defined, which has all the properties of the Euclidean one.[7] One important consequence of the theory is presented in section 14 (1871, p. 106): 'according as we impose elliptic, hyperbolic or parabolic geometry, the plane becomes a surface of constant positive, constant negative, or zero curvature'. This connects Klein's approach to Riemann's theory of the manifold. The important point for us is that, in the projective approach, manifolds with variable curvature are excluded from consideration from the outset.

Let me come back to Russell. As explained in Chapter 1, Russell came from the synthetic approach. He was thus reluctant to resume Klein's unrestricted use of numerical coordinates – he also refused the use of points with complex coordinates. This compelled him to make some adjustments. Russell gathered the hyperbolic and parabolic cases together (PoM, § 408) and reserved the elliptic case (where the Absolute is a complex curve) for separate treatment (PoM, § 408–9). The first theory is called the 'descriptive theory of distance', while the second one is named the 'projective theory of distance'.

First, a word on the 'descriptive' theory. The name comes from the fact that, as we have seen in Chapter 1, a descriptive space can be seen as a convex region of the projective space. Take the hyperbolic case. The real conic (the Absolute) 'divides' the real projective plane into two parts, one of which can be identified with the 'descriptive' plane. Then, by defining the distance between two points U and V of this plane as the logarithm between

U, V and the intersection A and B of the line (UV) to the conic, one ends up with a hyperbolic metric. In this plane, there will be an infinite number of lines passing through a point G not on Δ which do not intersect Δ (see Figure 2.2); and every point of the descriptive plane (inside the conic) will be infinitely distant from any points on the Absolute.

But one can also define a Euclidean plane by this method. Let us remove a line (the so-called line at infinity, the real branch of the degenerate imaginary absolute conic) from the projective plane and consider the restriction of the projective plane thus obtained. One can easily show that this restriction is a descriptive plane, and that the metric one obtains by applying Klein's definition is Euclidian. The distance between any point of the line Ω and any point of the descriptive space is then infinite, and there is only one parallel to a given line Δ passing through a point outside Δ (see Figure 2.3).

In order to deal with the elliptic linear distance, Russell, however, needs to consider the whole projective plane and not a proper restriction of it. This is why he said that the theory of elliptic distance is a 'projective theory'.[8] Following Klein, Russell asserts that, in this case, the Absolute is an imaginary conic. As he does not want to renounce his pure geometrical approach, he has then to give a geometrical interpretation of the complex points (PoM, p. 426):

But if such a function [the elliptic distance as the logarithm of a certain cross-ratio] is to be properly geometrical, it will be necessary to find

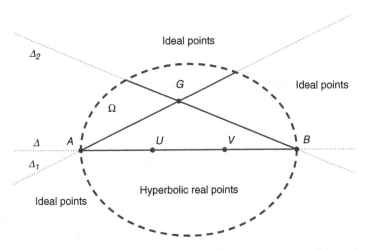

Figure 2.2 The Absolute is the real conic Ω. The two fundamental points of the line Δ are A and B. Δ_1 and Δ_2 are the 'ultra-parallel' of Δ passing through G. Compare with Figure 2.1

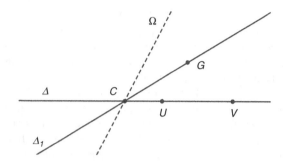

Figure 2.3   The Absolute is the line Ω. The fundamental point on the line Δ is C (the 'point at infinity'). Δ₁ is the only parallel to Δ passing through G. Compare with Figure 2.1

some geometrical entity to which our conjugate complex numbers ... correspond. This can be done by means of involutions.

The trick is a standard one. It was first stated by Von Staudt and maintained by Pasch and Klein. An involution I of points on a real projective line is a projective transformation of period 2 – that is, x and y being any given points on the line, if $I(x) = y$, then $I(y) = x$. In analytic geometry, an involution is expressed by a symmetric bilinear form. The set of the invariant points of the involution is thus either equal to a couple of real points (the hyperbolic case) or the empty set (the elliptic case).[9] In this case, the two roots of the quadratic form associated with the bilinear form are two imaginary conjugate numbers. Von Staudt's beautiful idea was to associate with each complex conjugate point pair the elliptic involution of the real projective line they determine.[10] Some niceties (which I will leave aside here)[11] are needed to develop this insight, but the essential thing is clear: conjugate point pairs can be placed in one to one correspondences with the elliptic involutions of the real projective lines, and can thus really be seen as involutions. This allowed Russell to accommodate Klein's analytic construction, without relying on imaginary points. Thus, for Russell, the Absolute of one-dimensional elliptic geometry is not an imaginary point pair, but an elliptic involution of the real line (and the Absolute of two-dimensional elliptic geometry is an elliptic polarity of the real plane).

Even if he slightly changed the original presentation, Russell maintained the key idea of Klein (1871): that the three classical metrical geometries deal with the properties of a certain quantity definable in projective terms. From this perspective, all the geometries dealing with spaces of constant curvature have the same status: they differ only in the choice of the Absolute. Hyperbolic and elliptic geometries were for the first time put on an equal footing with Euclidean theory.[12] The projective theory of metric introduced a sort of systematization in the (until then) chaotic field of non-Euclidean geometries, in the sense that the three cases were related to each other. Conics and cross-ratios were the key concepts on which the classification was based. All this was not new for Russell: PoM VI maintained on this point the (less rigorous) developments of *An Essay*.[13]

Musgrave has insisted on the fact that metrical geometries are mutually inconsistent and that the sole possibility of showing that they are parts of logic is to adopt the if-thenist stance. But in the projective approach, the alleged incompatibility between the classical metrical theories (hyperbolic, elliptic and Euclidean) is reduced to the differences between the properties of various kinds of conics. For Russell, then, saying that there are several different metrics is just as innocuous as saying that there are distinct projective figures (various sorts of conics). To take an analogy: in Euclidean geometry, one finds triangles and circles. No triangle is a circle. But this does not create any dramatic foundational problem related to the incompatibility between circles and triangles. In Klein's approach, the differences between the three classical metrical geometries boiled down to exactly this sort of difference – that is, to a difference between figures in a plane. And one must go even further. Klein prided himself on having succeeded in introducing unity and systematicity into the field of metrical theories.[14] He thus emphasized that Euclidean space is the limiting case between the hyperbolic and the elliptic space.[15] Figure 2.4 shows diagrams aimed at displaying how one can pass from an Absolute to another, and therefore at showing how the three classical metrical geometries are interrelated. From Klein's perspective, the alleged incompatibilities of the different metrical cases disappear to give place to a unified and articulated whole. Claiming with Musgrave that Russell's starting point in PoM was the incompatibilities between hyperbolic, elliptic and Euclidean geometries amounts to no less than passing over Klein's influence in silence – and to no less than seriously distorting Russell's thought.

There is another indirect way of showing that Musgrave's view is mistaken. As we have noted, Klein's theory excluded from consideration spaces with variable curvature. Now, if Musgrave were right, Russell could have used an if-thenist manoeuvre to resolve the incompatibility between theories of Riemannian manifolds with arbitrary curvature and the projective metrics. Provided that the geometries of space with variable curvature were axiomatizable, he could have held that the implication $T \Rightarrow p$ (with T the axioms of the theory and p a theorem) is a logical truth, and thus that these geometries belonged to logic. This is not what Russell did, however. Let me quote a passage from him (1902, p. 480):[16]

> If our coordinates are to represent any kind of spatial magnitudes, we must assume the possibility of equal quantities in different places, and hence, it will be found, we shall be compelled to regard the measure of curvature as constant. Let us examine the consequences of supposing it variable. In

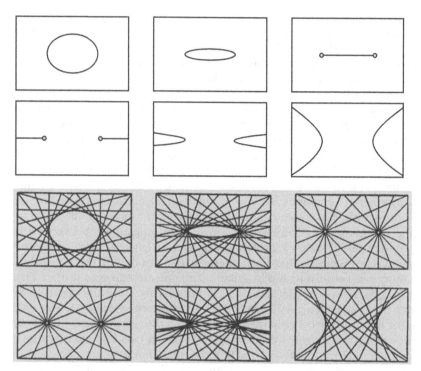

Figure 2.4   The transformation of the conic of equation $a_{11}u_1{}^2 + a_{22}u_2{}^2 - a_{33}u_3{}^2 = 0$

Note: The diagrams illustrate the transformation of the conic of equation (in homogeneous line coordinates) $a_{11}u_1{}^2 + a_{22}u_2{}^2 - a_{33}u_3{}^2 = 0$ (with a_{11}, a_{22} and a_{33} as positive and real), when a_{22} diminishes to become first equal to zero and then equal to a negative real number. The first series of six illustrations represent the conics as a set of points, while the six last ones adopt the dual point of view and represent the conics as a set of lines. Klein relates this process of deformation to the change of metric.

Source: Klein (1928, pp. 83–4).

the first place, the method of superposition would have become impossible, so that measurements could no longer be effected ... Thus a metrical coordinate system would become impossible. Moreover, geometry would become akin to geography; it would not consist of general theorems, but of descriptions of various localities. The variation of the space-constant would be not quantitative merely, but qualitative, and quantities in different places would be of different kinds. Thus the constancy of the measure of curvature is a precondition of any metrical coordinate system, and cannot be held doubtful while such a system is retained.

In 1902 (at that time, Russell had already written the essential parts of PoM VI), the concept of a space of variable curvature was still regarded as illegitimate. Only spaces with constant curvature, that is, spaces whose metrics can be introduced in the projective way, were accepted. Russell preferred to reject Riemannian geometries as illegitimate rather than accommodating them by using an if-thenist manoeuvre. This substantiates my view that the starting point of Russell's theory of metrical geometry in PoM was Klein's projective theory of distance, and not, as Musgrave has suggested, if-thenism and the mutual inconsistency between Euclidean and non-Euclidean geometries.[17]

2.3 Metrical geometry, distance and stretch

Even if it occupies a central place, the projective definition of metrical concepts is not the only characterization of distance and angle that Russell developed in PoM VI. In chapter 47, one finds two other ways of defining metrical concepts. The common point between these two paths is that, in each of them, metrical geometry is conceived as an independent science, which is not brought under the projective sway. That is, distance and angle are directly characterized, without resorting to any projective concepts. Now, the difference between the two is that in the first approach (which we will refer to here as the 'Leibnizian view'), presented in the first part of chapter 47, metrical geometry is regarded as a part of pure mathematics (i.e. of logic), while in the second one (which we will refer to as the 'divisibility view'), developed in the second part of the same chapter, metrical geometry is seen as an empirical science. I will present the two accounts in turn.

The Leibnizian view

In PoM §§ 392–5, Russell envisages the possibility of defining order and incidence between points, lines and planes in a real space in terms of distance. He first begins by characterizing the binary relation distance. In PoM § 392, he lists eight properties that a relation between points should have to be a distance.[18] Russell then raises two questions. The first is in § 394: is it possible to define order on descriptive lines in terms of distance? The second is in § 395: is it possible to define incidence from distance?

Concerning distance and order Russell writes (PoM, p. 409):

> It may be asked whether distances do not suffice for generating order on the straight line, without the need of any asymmetrical transitive relation of points. This represents, I think, the usual view of philosophers; but it is by no means easy to decide whether it represents a tenable view.

In descriptive geometry, the line is ordered by the betweenness relation. Now, let us imagine that a distance be defined on a three-dimensional real space (which is not further characterized by Russell) – would it be possible to derive from it the relation of order which holds between the points of the descriptive line? Russell answers positively to this question. But he emphasizes that, to achieve the task, the list of 'axioms' given in § 393 has to be considerably extended. Indeed, the main difficulty comes from the fact that distance is characterized there as a symmetrical relation.[19] This, explains Russell, makes the whole process of deriving betweenness from distance 'extremely complicated' (PoM, p. 410).

The more important issue concerning whether the concept of a line, and more generally the relations of incidence in the plane, can be derived from distance is broached in § 395, where Russell referred to the works of Pieri,[20] Leibniz, Frischauf and Peano[21] (PoM, pp. 410–11):

> Pieri has shown, in an admirable memoir, how to deduce metrical geometry by taking point and motion as the only indefinables ... The straight line joining two points is the class of points that are unchanged by a motion which leaves the two points fixed ... There is a method, invented by Leibniz and revived by Frischauf and Peano, in which distance alone is fundamental, and the straight line is defined by its means ... The locus of points equidistant from two fixed points is called a plane, and the intersection of two non-coincident *planes*, when it is not null, is called a *straight line*. (The definition of the straight line given by Peano is as follows: the straight line *ab* is the class of points *x* such that any point *y*, whose distances from *a* and *b* are respectively equal to the distances of *x* from *a* and *b*, must be coincident with *x*.) Leibniz, who invented this method, failed, according to Couturat, to prove that there are straight lines, or that a straight line is determined by any two of its points. Peano has not, so far as I am aware, succeeded in proving either of these propositions, but it is of course possible to introduce them by means of axioms ... In any case, however, the definitions prove that, by a sufficient use of axioms, it is possible to construct a geometry in which distance is fundamental, and the straight line derivative.

In section 2.2, we saw how Russell, in chapter 48, derived the metrical concepts from the projective notions. Russell was wondering here about the possibility of reversing the process. Indeed, if, starting from distance, lines and planes could be constructed and endowed with their usual descriptive relations, then, using Pasch's method, it would become possible to derive the whole of the projective theory. As Russell explicitly recognized, this path had been followed by many mathematicians and philosophers before him. It consists, at bottom, in taking distance and angle as the fundamental geometrical concepts. Thus, Poincaré, who grounded space on the notion of the isometric group, could be referred to as one of the supporters of this view.[22] Now, in § 395, Russell recognized that it is formally possible 'to construct a geometry in which distance is fundamental, and the straight line derivative'. He adds though that 'the method is so complicated as to be not practically desirable' (PoM, p. 411).

The 'Leibnizian' approaches are thus those which attempted to characterize descriptive (order) or projective (straight lines, incidence) notions in terms of the concept of distance, taken as an indefinable. What seems to characterize Russell's attitude towards Leibnizian views is that they are logically possible but highly artificial. That is, one can develop an axiomatic theory of metrical, descriptive and projective geometry which contains points, distance and angle as only indefinables (neither order, nor incidence);[23] but these systems would all be extremely complicated. Russell does not explain, however, what he meant by 'complicated'. From what he said, one can conjecture that an axiomatic system is complicated when one has to add many ad hoc conditions to get what one wants – when one is unable to control and anticipate the expansion of the list of postulates. But of course, this is not a clear and precise notion, and Russell does not say enough for us to clarify his thought. What seems to explain Russell's disdain for the 'Leibnizian' view is the importance he gave to the projective approach. Let me quote this telling judgement about Hilbert's *Grundlagen* from Russell's letter to Couturat dated 4 June 1904 (Schmid 2001, p. 403):

> [Hilbert] has gravely offended my aesthetic taste, for instance by his habit to discuss projective theorems within a Euclidean framework. He is not without merit, but the problems he deals with do not seem to me to be the most interesting ones.

The insight that projective theory must be placed at the foreground of the geometrical scene governs the whole of Russell's apprehension and seems to prevent him from attaching any philosophical importance to the Leibnizian attempt of defining the geometrical relations directly from the notion of distance and independently of the projective framework. Anyway, failing to provide a satisfactory explanation, let us retain Russell's twofold attitude: grounding geometry on metrical considerations is formally possible, but not philosophically interesting.

The divisibility view

In §§ 396–9, Russell turns to another conception, in which the projective framework is supposed as given, but in which the notion of a distance, instead of being defined in projective terms, is regarded as a new non-logical indefinable. Now, as I will explain more fully in Chapter 3, Russell considered projective geometry as a logical theory – and the projective theory of metric was thus regarded as purely logical knowledge. The idea that metrical geometry could be based on a new non-logical indefinable amounts then to asserting that metrical geometry, unlike projective geometry, should be regarded as an empirical science.

A few words, first, about terminology. In §§ 396–9, Russell heavily relied on his analysis of magnitude and quantity developed in PoM III (which we will examine in Chapter 4). There, he distinguished two different notions of measurable (extensive) magnitude: distance, defined as a binary relation with some particular formal properties; and magnitude of divisibility, defined as an empirical notion.[24] Until now, I have considered the metrical distance as a *distance* in the sense of PoM III (I put the word in italics when I specifically refer to the Russellian opposition between distance and magnitude of divisibility): in the projective theory, as in the Leibnizian view just considered, distance is defined as a relation, which fulfils the formal conditions expounded in PoM III. But there is another option: to consider the metrical distance, not as a *distance*, but as a magnitude of divisibility.

Now, what is a magnitude of divisibility? Russell's doctrine is very confused. Let me explain the general idea. Russell drew a distinction between quantities and magnitudes: quantities are the concrete items that have a certain magnitude. For instance two different bodies can have the same mass; the mass they have is their common magnitude, the bodies are the quantities. Magnitudes, explained Russell, are never divisible (PoM, p. 173), but it turns out that certain quantities can be divided into different parts, which are themselves quantities and have a magnitude. This is the case of bodies: one can divide a body into several parts, each of which has a mass. When this occurs, Russell claims that the quantities have 'magnitude of divisibility'. Three points must be kept in mind. First, it is not the case that all quantities have a magnitude of divisibility (pleasures are quantities which have magnitude, but which do not have a magnitude of divisibility). Second, magnitudes of divisibility have an additive structure. The possibility of dividing and concatenating quantities allows us to give sense to the addition of magnitudes. Thus, the mass of a body *c* composed of two bodies *a* and *b* would be equal to the mass of *a* plus the mass of *b*. Third, if the whole contains only a finite number of elements, then its divisibility is equal to its cardinality; if it contains an infinite number of elements, then the magnitude of divisibility still exists, but is much more complicated to define (for more on the theory of magnitude of divisibility, see Chapter 4).

In PoM III, Russell immediately applies his theory to the length of a segment (pp. 181–2). Two points *A* and *B* of the actual physical space determine a whole containing all the points between *A* and *B*. Russell calls this whole a 'stretch', and claims that it is 'a quantity, and has a divisibility' (ibid.). Now, according to him a stretch contains an infinite number of points. How, then, can we measure its divisibility? He answers thus (PoM, pp. 178–9):

> [In the case of infinite wholes] we require ... a method which does not go back to simple parts. In actual space, we have immediate judgements of equality as regards two infinite wholes. When we have such judgements, we can regard the sum of *n* equal wholes as *n* times each of

them; for addition of wholes does not demand their finitude. In this way numerical comparison of some pairs of wholes becomes possible. By the usual well-known methods, by continual subdivision and the method of limits, this is extended to all pairs of wholes which are such that immediate comparisons are possible. Without these immediate comparisons, which are necessary both logically and psychologically, nothing can be accomplished: we are always reduced in the last resort to the immediate judgement that our foot-rule has not greatly changed its size during measurement, and this judgement is prior to the results of physical science as to the extent to which bodies do actually change their size.

The possibility of measuring the magnitude of divisibility of a stretch is thus grounded on our capacity to perceive the equality in size of two segments. That a given stretch is twice as long as another one is not a mathematical or logical claim: it is an empirical truth.

Now that these preliminaries have been set down, let me come back to Russell's third theory of metrical geometry in PoM VI. The view is summarized in the following passage (p. 411):[25]

> We now start, as in descriptive geometry, with an asymmetrical transitive relation by which the straight line is both defined and shown to be a series. We define as the distance of two points *A* and *B* the magnitude of divisibility of the stretch from *A* to *B* or *B* to *A*.

In this approach, the linear[26] distance is no longer conceived as a formally characterized concept, but as an (empirical) magnitude of divisibility. Russell shows that, since divisibility has an additive structure, many properties that geometers attribute to metrical distance are satisfied by the divisibility of stretches. According to him, one would need only to set up three 'axioms'[27] to recover the expected properties of distance on the descriptive line. Note that he does not attempt here (as it was the case in the Leibnizian approaches) to ground descriptive (and projective) geometry on the theory of magnitude of divisibility. His ambition is more modest. He wants to show that, from within the descriptive (or projective) framework, it is possible to develop metrical (that is, hyperbolic, elliptic or Euclidean) geometry by introducing only the empirical concept of divisibility of a stretch. The notion is already endowed with nearly all the expected formal properties and, compared to the Leibnizian approaches, it is thus much 'simpler' to develop metrical geometries in this way (PoM, p. 414): 'the stretch in every series [descriptive line] is a quantity, and metrical geometry merely introduces such axioms as make all stretches of points measurable'.

Let me summarize the content of chapters 47 and 48 of PoM VI. There are three distinct theories of metrical geometries that one can distinguish by resorting to three different criteria. First, the view which identifies distance

Table 2.1   A summary of Russell's distinctions

	Metrical geometry as a pure logical science	Projective (descriptive) geometry required	Metrical geometry as an independent theory
The projective view	×	×	
The Leibnizian view	×		×
The divisibility view		×	×

with magnitude of divisibility is opposed to the two other approaches in that it is the only one which construes metrical geometry as an empirical discipline. Second, the Leibnizian approach stands against the other two, in that it is the only view which does not presuppose the projective (or descriptive) setting. Third, the projective theory of metric is the only view in which metrical geometry is defined as a sub-branch of the projective geometry. I have summarized the situation in Table 2.1.

The divisibility view requires the projective (or descriptive) setting, since the definition of a stretch refers to the concept of the projective (descriptive) line. On the other hand, Leibnizian approaches aim at basing the notion of space (and thus of projective and descriptive space) on a direct formal characterization of the notion of distance. At the same time, there is a distinction between the projective view and the divisibility approach. In the first conception, distance is a purely projective notion (a certain function of a certain cross-ratio), while in the latter, it is a perceptual relation (the divisibility of a stretch).

2.4   Poincaré and Russell on the perception of distance

Russell's claim, according to which one can empirically determine the metric of the surrounding space (which is linked to the idea that distance can be defined as a magnitude of divisibility), was the focal point of Poincaré's attacks in 1899. Russell was then aware of how disputable his assumption was. One might have expected that his doctrine of magnitude of divisibility provides us with some clues as to how he planned to meet Poincaré's challenge. But, as I have said, PoM's theory of divisibility is very muddled – so muddled that one nowhere finds an enumeration of the properties that the concatenation between quantities should have in order to be considered as having magnitude of divisibility.[28] This theory is thus not the right place to look at in order to understand Russell's view about Poincaré's conventionalism. In this somewhat digressive section, I would like to examine in more detail the idea that one can perceive distance.

The main argument Poincaré used to support his view that all our measures apply to bodies, and not directly to space, so that any seemingly non-Euclidean result can be interpreted as the result of the physical deformation of the bodies (both the body measured and the measuring ruler) used in the measurement process. This insight is captured well in Poincaré's notorious description of the non-Euclidean world (1902, pp. 88–91). It is important to understand that Russell did not reject this argument. Criticizing the Euclidean concept of superposition, he thus wrote (PoM, p. 405):

> To speak of motion implies that our triangles are not spatial, but material. For a point of space is a position, and can no more change its position than the leopard can change his spots ... Motion, in the ordinary sense, is only possible to matter not to space. But in this case superposition proves no geometrical property. Suppose that the triangle *ABC* is by the window, and the side *AB* consists of the column of mercury in a thermometer; suppose also that *DEF* is by the fire. Let us apply *ABC* to *DEF* as Euclid directs, and let *AB* just cover *DE*. Then we are to conclude that *ABC* and *DEF*, before the motion, were equal in all respects. But if we had brought *DEF* to *ABC*, no such result would have followed. But how foolish! I shall be told; of course *ABC* and *DEF* are to be both rigid bodies. Well and good. But two little difficulties remain. In the first place – and for my opponent, who is an empirical philosopher, this point is serious – it is as certain as anything can be that there are no rigid bodies in the universe. In the second place – and if my opponent were not an empiricist, he would find this objection far more fatal – the meaning of rigidity presupposes a purely spatial metrical equality, logically independent of matter.

That the 'meaning of rigidity presupposes a purely spatial metrical equality' is exactly Poincaré's main claim. The concept of a rigid or solid body, far from being an empirical representation, is a concept defined by the mathematicians (Poincaré 1902, p. 93):

> Geometry would be only the study of the movements of solid bodies; but, in reality, it is not concerned with natural solids, its object is certain ideal solids, absolutely invariable, which are but a greatly simplified and remote image of them. The concept of these ideal bodies is entirely mental, and experiment is the opportunity which enables us to reach the idea. The object of geometry is the study of a particular 'group'; but the general concept of a group pre-exists in our mind, at least potentially. It is imposed on us not as a form of our sensitiveness, but as a form of our understanding; only, from among all possible groups, we must choose one that will be the standard, so to speak, to which we shall refer natural phenomena.

Reference to the concept of 'group' aside, Russell follows quite faithfully the line developed here by Poincaré. The notion of a solid body is not, for Russell, an empirical concept.[29]

Poincaré concluded from this argument that metrical geometry was not an empirical science. Russell refused to follow him on this. But how could he? In the review he devoted to the English translation of *La Science et l'Hypothèse*, Russell tackled this problem (1905a, p. 591):[30]

> The second part, on Space, repeats the contention that none of the various Euclidean and non-Euclidean geometries is truer than another, but that the Euclidean is the most convenient. The argument is, that all our experiments concern *bodies*, and that any apparently non-Euclidean result can be interpreted as due to the nature of bodies, not to the nature of space. Admitting this, I do not think the consequence follows ... There are relations which arrange the points of space in any order imaginable, e.g. so that objects which we perceive as near together would be widely separated, while objects which, in the perceived spatial order, are very distant, would come between objects which are very near to us. In short, relations subsist between points which make a complete re-arrangement of them, not at all resembling the arrangement we perceive. These other arrangements differ from the one we perceive, it would seem, just in the fact that we do not perceive them; and this brings out the necessity of supposing that the spatial relations we regard as actual *are* perceived.

After having 'admitted' Poincaré's thesis that one cannot measure space, Russell refused to endorse what was viewed by Poincaré as a direct consequence of it: none of the various metrical geometries is truer than another. Russell's alternative suggestion is that measurement might not be the sole possible way to determine empirically what the 'true' metric is. For instance, people could have an immediate perception of distance. This is consonant with what Russell said in PoM III (p. 178): 'in actual space, we have immediate judgements of equality as regards two infinite wholes'. That two stretches a and b have the same magnitude of divisibility does not follow from an operation of superposition of two segments, but from the simultaneous perception of a and of b. Poincaré's conventionalist conclusion follows from the assertion that one can measure only bodies and not portions of space, only if one maintains that the judgement of equality in length between two stretches is grounded on superposition and measurement. But for Russell, this is not the case: measurement is not the only source of metrical judgements. To understand the real philosophical import of Russell's move, it is necessary to broaden the perspective and to come back to Kant's theory of perception.

As is well-known, Kant distinguished between two components in empirical intuition: the sensation proper (the matter of intuition) on the one hand, and space and time (the forms of intuition) on the other. Kant emphasized that 'that within which the sensations can alone be ordered and placed in a certain form cannot itself be in turn sensation' (1781, p. 156).

For him, space was then not a part of the matter of the sensation. Poincaré shared much of Kant's framework. He especially maintained his 'atomistic' view of perception, according to which sensations were like atoms united in a form, which could not itself be sensed.[31] Thus (1902, p. 58): 'none of our sensations, if isolated, could have brought us to the concept of space; we are brought to it solely by studying the laws by which those sensations succeed one another'. However, he, unlike Kant, did not think that space was a form of our sensibility – for him, space was a concept, namely the notion of a transformation group (to which refers the phrase 'the laws by which those sensations succeed one another' in the passage just quoted), freely created by the mathematician. What is important is nevertheless the fact that, for him as for Kant, there is no sensation of space. Space is a form, not a matter of sensation.

Russell's idea that 'we have immediate judgements of equality as regards two infinite wholes' amounts to no less than challenging the Kantian atomistic view of perception. This is explicitly contended in a letter to Couturat dated 4 April 1904 (Schmid 2001, p. 377):

> [Here is the important point:] it is relations, and not points, that characterize actual space. It is therefore necessary to admit that sensation – or perception, if you prefer – reveals relations as much as the terms of relations. All kinds of relations subsist between the points of actual space; but there is only one kind of relation that we perceive immediately. The immediate perception of relations is (if I am not mistaken) something that Kant does not admit. But here is a case in which this perception is evident.[32]

Russell dropped here any reference to magnitude of divisibility. But he continued to claim that we can directly perceive metrical relation between points of the actual space, without passing through the mediation of measurement. What is new is the acknowledgement that this assumption, since it leads to an admission that sensation or perception 'reveals relations as much as the terms of relations', goes directly against Kant's atomistic view of perception.[33] It is likely that in 1904 Russell connected this insight to Meinong's radical anti-Kantian approach. At the time, Russell was indeed reviewing *Über Annahme*, where Meinong developed his criticism of the atomistic view of perception. Meinong maintained that we can have a perception of a 'higher-order object', and, in his review,[34] Russell resumed the standard example of the perception of melody (which is for Meinong a relation): the apprehension of a melody is clearly grounded on the perception of the various notes it is composed of, but the sensation of the melody is neither reducible to the perception of a note, nor a mere unperceivable form relating the different sensations.[35] This pre-Gestaltist conception runs clearly against the Kantian tradition, since it claims that a certain sort of relation between sensations can be perceived. This Meinongian outlook was certainly at the back of Russell's mind when he wrote to Couturat. Distance, as a relation between points, can be conceived as a 'higher-order object', as a kind of Gestalt, that we can apprehend when we simultaneously look at two stretches.

Thus, to conclude, it seems that the opposition between Russell and Poincaré concerning the empirical status of metrical geometry could be traced back, at least partially, to a difference between their respective conceptions of perception. Poincaré, following an atomistic approach of sensation, maintained that one could not have a direct perception of distance, and that the sole way to determine a distance was to measure it. Russell shared Poincaré's view according to which the notion of a rigid body is not empirical, but he believed that one could make (arguably crude) immediate empirical judgements concerning the size of two segments, and that these perceptual judgements were not grounded on a measurement process. This allowed Russell to refuse (as did Poincaré) that one could measure the curvature of our space, while maintaining (unlike Poincaré) that such a curvature could nevertheless, in principle (that is, leaving aside all the problems posed by the accuracy of our perception), be empirically determined – direct perception (if not measure) can give access to metrical facts. Of course, it is difficult to understand more fully what Russell was aiming at: how exactly are we supposed to determine the curvature of our space without resorting to any operation of measurement? To perceive directly certain facts about segment sizes is one thing – to extract from these judgements enough information to draw a conclusion about the metric of our space is another completely different matter. It seems however that Russell did not hold[36] that we can have decisive evidence about the metric of our space; his aim, as we will soon see, was rather to defend the weaker claim that the notion of a distance could have an empirical origin.

However problematic Russell's empirical conception of distance might have been, the idea that Poincaré's conventionalism presupposes an atomistic conception of perception is attractive and valuable. It seems that the French mathematician modelled his psychological genesis of the concept of space on the process of measurement. Indeed, as is well known, the distinction between a change of states and a change of positions is for Poincaré crucial. A change of position is a modification of the aggregate of impressions such that 'we could restore the primitive [situation] by making movements which would confront us with the same movable object in the same relative position' (1902, p. 58). This possibility of compensating a change, which is at the source of the notion of inverse operation (and thus of a group), is also a feature which is intimately connected to measurement. A process of measurement can be seen as a way to 'restore' a change: the distance measured indicates how much one should move the ruler to put the two extremities of a segment in the same relative position to the origin.

This link between the working of human perception and the measurement process is acknowledged by Poincaré (1908, p. 89):[37] 'we would not have been able to construct space if we had not had an instrument for measuring it; well then, our own body is the instrument to which all is referred, and of which we instinctively make use'.

For Poincaré, people do not sense space. They have however the capacity to compensate certain sensorial modifications (in a sense, to 'measure' them), and that is why they are led to set up the distinction between change of positions and change of states – the distinction which is at the root of the construction of the concept of space. Russell is then right when he asserts that Poincaré's conventionalism does not proceed only from the idea that the notion of rigidity presupposes the definition of a metric, but also from the further assumption that measurement is our only access to distance and space. For Poincaré, perceiving is nothing but measuring (our body is a measuring instrument). And it seems to me that Russell is right too when he suggests that this theory of perception is not forced on us.

2.5 Metrical geometry as an empirical science

Let me come back to my main line of interest. In PoM, Russell developed three conceptions of metrical geometries: the projective view, where distance is derived from cross-ratio; the Leibnizian view, where distance is conceived as a formal relation, axiomatically characterized; and the divisibility view, where distance is seen as a magnitude (of divisibility), empirically given. The two first approaches construed metrical geometry as a pure mathematical science (metric is there a logical concept), while the last one conferred on metrical geometry the status of an empirical science. Among these three formally perfect characterizations, which definition is the most fundamental one? Russell's answer is summarized in the last page of chapter 48 (PoM, p. 428):

> To sum up: although the usual so-called projective theory of distance, both in descriptive and in projective space, is purely technical, yet such spaces do necessarily possess metrical properties, which can be defined and deduced without new indefinables or indemonstrables. But metrical geometry, as an independent subject, requires the new idea of the magnitude of divisibility of a series, which is indefinable, and does not belong, properly speaking, to mathematics ... Thus there is a genuinely distinct science of metrical geometry, but, since it introduces a new indefinable, it does not belong to pure mathematics in the sense in which we have used the word in this work ... On the other hand, projective and descriptive geometry are both independent of all metrical assumptions, and allow the development of metrical properties out of themselves; hence, since these subjects belong to pure mathematics, the pure mathematician should adopt their theory of metrical matters. There is, it is true, another

metrical geometry, which does work with distances, defined as one–one relations having certain properties, and this subject is part of pure mathematics; but it is terribly complicated, and requires a bewildering number of axioms. Hence the deduction of metrical properties from the definition of a projective or a descriptive space has real importance, and, in spite of appearances to the contrary, it affords, from the point of view of pure mathematics, a genuine simplification and unification of method.

Unsurprisingly, Russell makes it clear that, among the two mathematical (non-empirical) conceptions of metric, the projective view has his preference. He believes that metrical geometry can be developed along Leibnizian lines, but this possibility is only formal and devoid of any philosophical importance. From our perspective, the most interesting comparison is however the one Russell is making between the projective and the empirical approach. Indeed, he clearly favoured the latter conception: 'the usual so-called projective theory of distance ... is purely technical', 'metrical geometry, as an independent subject, requires the new idea of magnitude of divisibility of a series', 'there is a genuinely distinct science of metrical geometry, but since it introduces a new indefinable, it does not belong to pure mathematics in the sense in which we have used the word in this work'. Russell is claiming here that metrical geometry, as an independent science, must be construed as an empirical discipline. This is surprising. As we will see in Chapter 3, projective geometry is considered to be a logical science by him. In Klein's view, metrical geometry is just a subfield of the projective theory – it must be counted then as a logical science as well. In other words, no technical obstacle forbad integrating metrical geometry into the sphere of logical science. And yet, Russell chose to expel metrical geometry from pure mathematics. Why did he renounce such an easy victory? Why did he finally consent to introduce distance as a new indefinable?[38]

One could at first suggest the following answer: that there is a perception of length is a given datum that every analysis of metrical knowledge has to take into account. So, even if a pure logical theory of distance were to be developed, it could not be taken as the last word on the issue. If everyone were to agree that distances are perceived, the projective derivation of metric would be regarded (at best) as a mathematical description of an empirical concept, but it could not be viewed as a definition of a logical notion. But as a matter of fact, not everyone at the beginning of the twentieth century thought that metrical notions were abstracted from perception. Poincaré, for instance, developed a bunch of influential arguments (some of which were shared by Russell) aimed at showing that the notion of a distance was not derived from experience. This first answer does not work, and, in fact, one does not find it in Russell's writings. In PoM VI, Russell does not ground his idea that metrical geometry is ultimately an empirical science on the fact that people perceive distance.

To understand Russell's actual reasoning, one should pay attention to the distinction between metrical geometry as a *part* of the projective theory, and metrical geometry as an *independent* subject, as a *genuinely distinct* science. Why exactly did Russell not want to define distance as a logarithm of a cross-ratio? Because this characterization makes metrical geometry appear as a marginal part of the projective theory. Let me quote him writing of the projective definition of distance (PoM, p. 425):

> It may well be asked, however, why we should desire to define a function of two variable points possessing these properties. If the mathematician replies that his only object is amusement, his procedure will be logically irreproachable, but extremely frivolous. He will however scarcely make this reply. We have, as a matter of fact, the notion of a stretch, and ... we know that the stretch has magnitude ... Thus the descriptive theory of distance, unless we regard it as purely frivolous, does not dispense with the need of [the theory of the magnitude of divisibility] ... The stretch, in descriptive space, is completely defined by its end-points, and in no way requires a reference to two further ideal points.

Russell is even clearer two pages later, in the conclusion of his discussion of the projective theory of elliptic metric (PoM, p. 427):

> But to [the projective theory of distance], as anything more than a technical development, there are the same objections as in the case of descriptive space; i.e., unless there be some magnitude determined by every actual point-pair, there is no reason for the process by which we obtain the above measure of distance; and if there is such a magnitude, then the above process gives merely the measure, not the definition, of the magnitude in question. Thus the stretch or distance remains a fundamental entity, of which the properties are such that the above method gives a measure of it, but not a definition.

The projective approach of metric gives 'no reason for the process by which we obtain the ... measure of distance'. That this is indeed the case is clearly illustrated by Klein's actual construction. If 'logically irreproachable', Klein's derivation is based on many ad hoc decisions.[39] Thus, Klein restricted the values of certain parameters to guarantee that the distances remained real-valued;[40] he also did not consider the distances between all the projective points (for instance, the properties of the distances between two points separated from each other by the two fundamental points in the hyperbolic case are not studied);[41] finally, the 'special' case of the Euclidean metric required many 'adjustments' to recover the expected formula.[42] From Klein's perspective, these various decisions were explicitly taken to recover the basic properties of the notion of distance, which was thus regarded as being already given. If the projective derivation unified the field of metrical geometries, the way the particularizations of the projective common setting were elaborated did not come from projective geometry itself. Russell was right then: Klein's conception presupposed the concept of a distance. This was not a defect, since Klein, far from willing to develop the metrical properties out of the projective setting, only wanted to display how the projective tool enabled us to uncover the deep unity underlying the different classical metrical geometries. But to transform this mathematical achievement into a philosophical analysis of distance would however be a mistake.

Ironically, Russell completely agreed with Poincaré's diagnosis concerning Klein's work. Poincaré criticized the projective conception of metric (he called it 'Von Staudt's theory') in which he saw a possible alternative to his own group-theoretical view (1898, pp. 35–7):

> This ... is the weak point of [Von Staudt's] theory, attractive though it be. To arrive at the notion of a length by regarding it merely as a particular case of the anharmonic ratio is an artificial and repugnant detour. This evidently is not the manner in which our geometrical notions were formed ... If we had proceeded, as the geometry of Von Staudt supposes us to have done, some Apollonius would have discovered the properties of polar. But it would have been long after that the progress of science would have made clear what a length or an angle is. We should have to wait for some Newton to discover the various cases of the equality of triangles. And this is obviously not the way that things have come to pass.

In the projective approach, metrical theory seems to be a frivolous invention of some mathematical genius, a marginal corner of the all-encompassing projective theory. Now, for Poincaré, there is no doubt that metrical geometry is an independent science – the history of mathematics shows that the various cases of triangle equality are more fundamental than the complicated development around projective cross-ratios. As Von Staudt's (Klein's) derivation of metric cannot account for the centrality and the independence of metrical geometry in the mathematical sciences, it must be rejected.

Russell, like Poincaré, endorsed the view that the projective derivation does not grant metrical geometry the status it deserves. Klein's reasoning is 'purely technical': it provides us with a 'measure', and not with a 'definition', of a distance. But Russell did not adopt Poincaré's Leibnizian view of metric – he did not want to install the metrical space at the centre of the geometrical scene. For the reasons we have enumerated in the previous chapter, Russell remained faithful to the Von Staudtian insight that projective space is the fundamental geometrical concept. Owing to this, the only remaining solution was to split metrical geometry into two distinct disciplines. As a purely mathematical theory, metrical geometry can be developed as a subfield of projective geometry. But this way of looking at the field does not explain why metrical geometry enjoys the place it has in the architecture of the sciences. To account for the independence and centrality of the metrical concepts, one should refer to a new extra-logical indefinable, the divisibility of the stretch. The situation is then tricky. Thanks to Klein, the Russellian logicist can account for the content of metrical geometry – and in this way, metrical geometry can be counted as a part of pure mathematics. But the logicist cannot account for the role metrical geometry plays in science – and, in this respect, he or she should renounce viewing this discipline as an integral part of logic. It should be highlighted that, in this reading, the importance given to the empirical characterization of metrical geometry in PoM VI does not come from the conviction that people can perceive distance. On the contrary, it is because he wanted to account for the independence of metrical geometry that Russell resolved to introduce a new non-logical indefinable, and endowed people with the capacity to perceive distance. This is thus an insight about the architecture of scientific building, and not an epistemological consideration about human perception, which is at the basis of Russell's choice in favour of the empirical approach. This conclusion fits well with what we have explained above: Russell did not claim that one can experimentally determine the curvature of our space; he just emphasized that Poincaré's conventionalism depended on an epistemological assumption about human perception which is not forced on us.

Let me come back, once again, to Musgrave. This scholar based his reconstruction on two assumptions: first, that Russell's central problem was to account for the fact that the various metrical geometries were mutually incompatible; second, that Russell relied on if-thenism to explain the articulation between pure and applied geometry. We have seen above (section 2.2) what can be thought about the former thesis. Concerning the distinction between pure and applied geometry, it is worth emphasizing that the relationship between the empirical and the projective metric does not follow the if-thenist schema. According to my reading, Russell introduced the idea that there is an empirical metrical geometry because the projective theory could not give to the metrical geometry the status it deserves. To grant metrical geometry its independence, Russell had to break with the conceptual scaffolding of the projective approach – he had to reorganize the conceptual structure of the doctrine: distance, which was regarded as a function of a complicated non-primitive notion (cross-ratio), becomes an indefinable. In other words, the metrical theory, considered as a doctrine of magnitude of divisibility, and the metrical theory, viewed as a theory of a certain kind of function of cross-ratio, do not have the same structure at all. The primitive propositions of the former are not the 'protasis' (T) of the 'hypotheticals' (T$\Rightarrow$p) that compose the latter. The passage from the pure projective theory to the empirical metrical geometry does not correspond to a simple 'discharge' of the antecedents of the conditionals. What is wrong in the if-thenist perspective is that the relationship between pure and empirical metrical geometry is reduced to a mere formal trick – the applied theory shares exactly the same conceptual form as the pure one, only the status of the antecedents of the conditional differs. But PoM VI does not work this way. To account for the independence of the metrical theory, Russell had to redesign completely the conceptual structure of the doctrine of projective distance. Musgrave's reading misses this important point.

A puzzle remained, however. As we saw, Russell had the technical means to include metrical geometry in the sphere of the logical sciences. He chose not to do so because such a manoeuvre would construe metrical geometry as a marginal corner of the projective theory. But it is not easy to understand why he gave so much weight to these considerations about the architecture of science. Klein provided him with the means to derive all the metrical theorems from projective geometry (and thus from logic – see pp. 83–4). Why not then maintain that metrical geometry is really a logical science, and explain the place it has in the sciences by the way people have come to know this content? As we will see in the next chapter, Russell was prone to distinguish between the mathematical content and the psychology of the mathematicians (that is, the various ways mathematicians come to know mathematical truths). Why did he not use this distinction to back up Klein's projective theory of metric? There is no doubt that metrical geometry played a central role in the history of mathematics. But Russell could have regarded this historical fact as something which has nothing to do with the genuine content of metrical geometry. Instead of doing so, he maintained that metrical geometry should be regarded as an independent discipline, and that it should be expelled from the logical sphere. What needs to be clarified is the reason he had for believing that the architectonic features of metrical geometry should be seen as an integral part of their content – the reason why he preferred restricting the scope of his logicism (metrical geometry is not a pure logical science) to distorting the internal organization of the mathematical sciences (metrical geometry must keep its central position).[43]

2.6   Conclusion

The conception of geometry Russell developed in PoM VI was deeply rooted in a particular tradition of geometrical thought, which was well represented in the second half of the nineteenth century. This does not mean that Russell's view follows directly from the development of the mathematical researches of his time. To root his thought in a very particular tradition could have the drawback of closing one's mind to certain other possibilities. Thus, as Torretti showed (1978, pp. 314–18), Russell does not seem to have understood Gauss's idea of intrinsic geometry, and he missed the strength and depth of Riemann's works. Russell also referred only superficially to

Lie's and Poincaré's approach of geometry in terms of continuous groups. Last but not least, I am not sure that Russell grasped the novelty of Hilbert's axiomatic works (see the passage quoted on p. 63). One can criticize Russell for not having been fully aware of the importance and scope of these major contributions. My primary aim here was however to show that these omissions are the counterparts of a deep involvement in the projective tradition – an involvement that still had mathematical sense at the time. It would then be a mistake to read the opposition between Russell on the one hand and Poincaré, Riemann and Hilbert on the other hand as an opposition between the 'Beotians'[44] (who knew nothing about the ongoing geometrical researches) and the working mathematicians.[45]

When Russell came to the question of metric, the projective pedigree of his thought posed a great difficulty for him. When one adopts Klein's projective characterization, one loses the fact that metrical geometry is an independent mathematical branch. Russell had then three possibilities. He could firstly renounce giving a logical account of the centrality of metrical concepts in the sciences, arguing that this feature is not a part of the mathematical content of these notions. For some still unclear reasons, Russell did not take this path. He could secondly throw back into question the idea that, in pure mathematics, projective geometry is the basic theory. But Russell did not retain this option and he systematically downplayed the Leibnizian attempt deriving the projective theory from the formal definition of a distance. The only remaining option was to present metrical geometry as a combination of two components: a projective (or descriptive) framework and the properties of a new non-logical constant, the magnitude of divisibility of the stretches. This is the path that Russell took in PoM VI. It had the advantage of keeping projective (descriptive) space at the centre of the picture, while preserving the independence of metrical geometry. But there is a price to pay: metrical geometry is ultimately seen as a non-logical science.

In the preceding chapter, I highlighted that Russell, far from contenting himself with finding one formally adequate analysis of projective geometry, attempted to discover the true analysis of projective space. The situation concerning metrical geometry illustrates once again that Russell's problem is an embarrassment of riches. The logicist developed three ways of defining the metrical space; but he needed to end up with only one analysis. How then could he make his choice? The situation is however more dramatic here than in the projective case, since he chose to restrict the scope of his logicism rather than sacrificing the independence and autonomy of metrical geometry. This point is important. Scholars of Russell, because they usually focus on his long battle with the paradox, seem to presuppose that the only reason for a logicist not to include a part of mathematics in logic is 'technical': a piece of mathematics should be regarded as external to logic only if one could not derive it from logical principles and logical notions.

Thus, because of the paradoxes, Russell had been led to abandon the idea of showing that the number of entities is not finite.[46] PoM VI shows that the situation is more complicated, and that one could also possess 'non-technical' reasons to refuse to reduce a given piece of mathematics to logic. Russell had the logical means to develop a purely logical theory of metrical space; but he did not, for the reasons we have seen. This case is thus particularly interesting, since it vividly shows that the non-formal constraints put on the logicization can run counter to the opportunities offered by the logical toolkit – and that logicism is not only a technical project. It remains, however, to clarify the nature of these non-formal constraints.

3
Geometry, Logicism and 'If-Thenism'

In the two previous chapters, we have dealt in some detail with Russell's view of projective and metric geometry. But we have not yet explained how he viewed the articulation between geometry and logic. Russell is well known for having claimed that geometry (projective geometry at least) is a part of pure mathematics. But how did he proceed to reduce geometry to logic? As we began to see in Chapter 2, the standard answer is 'if-thenism': to regard a given theory as a part of logic, it is sufficient to show that the said theory is consistent (has a model) and can be axiomatized.[1] This interpretation confronts us with a problem however. Russell included projective geometry within pure mathematics, but he did not believe that metrical geometry was a logical science. For all that, he admitted that metrical geometry was consistent and could be axiomatized – in at least two different ways: via the projective definition of metric and also in the direct Leibnizian way. Axiomatization and consistency cannot then be the sole criteria Russell used to characterize the sphere of logical science. More is needed – but what exactly?

The question concerning the distinction between the empirical and logical sciences is in reality only a variation on a more general theme. As we saw in the two last chapters, Russell, in PoM VI, never stopped facing multiple reduction problems. There was always more than one way to define the basic geometrical concepts, and the main issue was to choose the 'good' ones. How did Russell proceed to solve this problem? As for determining whether a given science was logical or not, we saw that, to sort out the different possible alternatives, he was forced to use non-formal criteria. This prominence of non-formal considerations raised a host of new questions. In this chapter, I will begin to explore these issues, independently of their geometrical context – and I will resume this analysis in Chapters 6 and 7.

I start my analysis with a discussion of if-thenism. In sections 1 and 2, I show that, contrary to what the if-thenists suggest, Russell never provided us with a formal recipe allowing us to classify a given branch of knowledge as a logical theory. Axiomatization and consistency are necessary conditions for a given science to be logical; but these conditions are not sufficient, and the issue whether or not a given body of knowledge is logical or empirical remains a question to be solved on a case-by-case basis. In section 3, I draw attention to the potential conflict between the role played by extra-formal considerations in Russell's analytical practices and his outward display of anti-psychologism. Any topic-specific feature Russell wanted to preserve in his logical analysis can be suspected of being a psychological ingredient which should be carefully severed from the mathematical content. In section 4, I focus on the notion of a relational type Russell brought forward at the beginning of PoM. This neglected concept provided Russell with a way of giving a logical status to the topic specificities of the existing mathematical branch, and to save his approach from the psychologist collapse. But the tension resurfaced at another level. The concept of a relational type merges two distinct components, a logical one (types are kinds of relation logically characterized) and a historical one (relational types are at the basis of some branches of existing mathematics). But Russell nowhere provides us with the means to control the balance between these two elements.

3.1 Russell's if-thenism (I): definitions

Russell, when describing his logicist programme, used to differentiate between two tasks. For instance, in § 474, at the extreme end of PoM, he wrote (p. 497):

> [In the process of deriving pure mathematics from the logical indefinables], two points are specially important: the definitions and the existence-theorems. A definition is always either the definition of a class, or the definition of the single member of a unit class: this is a necessary result of the plain fact that a definition can only be effected by assigning a property of the object or objects to be defined, i.e., by stating a propositional function which they are to satisfy. A kind of grammar controls definitions, making it impossible e.g. to define Euclidean *Space*, but possible to define the class of Euclidean *spaces* ... The existence-theorems of mathematics – i.e. the proofs that the various classes defined are not null – are almost all obtained from Arithmetic. It may be well here to collect the more important of them.

For reducing mathematics to logic, it is then necessary, firstly, to define the various mathematical objects in terms of logical constants, and, secondly, to prove, by using only logical principles, that the concepts thus defined have an instance. In this section and the following one, I will examine these two demands in turn. My aim is to determine whether the necessary conditions listed are also sufficient to delimit the sphere of the logical sciences.

First, what counts as a logical definition in PoM? It is well known that Russell held in PoM that only nominal definitions are logically admissible, and that definition by postulates must be rejected (p. 112). It is important, however, not to be misled by this claim. I will explain the point by taking the example of Russell's well-known discussion of Peano's arithmetic in PoM II. Peano's axiomatic system is composed of five postulates and contains three indefinables: '0', 'number' and 'successor' (1903, p. 125):

(1) 0 is a number. (2) If *a* is a number, the successor of *a* is a number. (3) If two numbers have the same successor, the two numbers are identical. (4) 0 is not the successor of any number. (5) If *s* be a class to which belongs 0 and also the successor of every number belonging to *s*, then every number belongs to *s*.

Peano acknowledged that 'there is an infinity of systems satisfying all the primitive propositions', but that 'all the systems which satisfy the primitive propositions have a one–one correspondence with the numbers' (ibid.). This one–one correspondence was indeed at the basis of Peano's characterization of ℕ: 'number is what is obtained from all these systems by abstraction; in other words, number is the system which has all the properties enunciated in the primitive proposition, and those only' (ibid.). One could thus say that Peano endorsed a sort of structuralism: according to him, whole numbers were the bare places of the structure implicitly defined by the axioms, and exemplified by all the 'models' of his (second-order) theory. What did Russell criticize in this approach?

It is sometimes suggested that Russell objected to the very idea of an implicit definition – for him, Peano's characterization would be wrong because it would not succeed in defining anything. This interpretation is false. Russell does not say that Peano's system does not define anything, but that the concept that it defined is not the one one seeks to characterize. Peano's definition by postulates is a nominal (i.e. correct) definition of a certain structure, called by Russell a 'progression'. A Russellian progression corresponds to our *ω*-sequence – it is an ordered set isomorphic to ℕ. According to Russell then, Peano's mistake was not to have used a wrong process of definition, but to have confused the general notion of a progression with the particular progression that ℕ is. To show the distinction between progression and natural numbers, Russell suggests replacing 0, number and succession with three distinct variables. Once such a substitution is made, Peano's postulates appear as a nominal definition of a propositional function with three variables, which could be applied, among other things, to the structure ℕ (PoM, p. 126):

It is very easy so to state the matter that the five primitive propositions become transformed into the nominal definition of a certain class of trios. There are then no longer any indefinables or indemonstrables in our theory, which has become a pure piece of Logic. But 0, number and succession become variables, since they are only determined as one of the class of trios.

This move can be generalized. According to Russell, any implicit definition (any axiomatic characterization)[2] can be changed into a nominal definition by transforming the non-logical constants into variables (of the suitable types), and by considering that the axiomatic characterizes the class of its models.

In the extract from PoM § 474 quoted above, the logical definitions which Russell referred to are precisely the nominalized axiomatic definitions we are talking about. Russell took there the example of Euclidean space. From the usual (let us say Hilbertian) axiomatic, one can form a nominal definition of the class of Euclidean spaces. An analogue of Peano's mistake could then arise – it would consist in believing that the Euclidean axioms define a single Euclidean space, while they characterize in reality a class of relational structures.[3] The conjunction of Euclidean axioms, like the conjunction of Peano's axioms for arithmetic, once turned into a nominal definition, is a second-order propositional function which can be applied to all the sets of first-order properties and relations which fulfil the conditions described in the postulates. Thus, Russell does not say that implicit definitions are wrong – he only claims that they could lead us to believe mistakenly that what is defined is one particular relational structure, while it is actually a class of relational structures (a second-order concept).[4]

We should not let ourselves be mesmerized by Russell's attack on definition by postulates. He has nothing against the idea that an axiomatic system defines something – he only stresses the fact that what is defined is not the thing that it is sometimes believed to be. To summarize my point, let me quote the introduction in Whitehead (1906b, pp. 1–2):

> Geometry, in the sense in which it is here considered, is a part of Pure Mathematics, and like such sciences it is composed of Definitions, Axioms, Existence-theorems, and Deductions ... The geometrical axioms are statements about relations between points; but they are not statements about particular relations between particular points ... Accordingly – since the class of points is undetermined – the axioms are not propositions at all: they are propositional functions ... Some authors term the axioms 'definitions' of the undetermined entities to which they refer. The enunciation of axioms is then said to be the process of 'definition by postulates'. There is no objection to this phraseology, so long as it is clearly understood that in general – and certainly in Geometry – the axioms do not characterize one unique class of entities (the points); but that many ... determinations of the class of points are possible, consistently with the truth of the axioms.

From a logicist point of view, 'there is no objection to [the phraseology of definition by postulate], so long as it is clearly understood that in general ... the axioms do not characterize one unique class of entities (the points)'.

Let me now come back to the case of the logical definition of space in PoM. The following passage is often quoted by the if-thenists (pp. 429–30):

> In the previous chapters, I have spoken, as most authors do, of certain indefinables in Geometry. But this was a concession, and must now be rectified. In mathematics, two classes of entities which have internal relations of the same logical type are equivalent. Hence we are never dealing with one particular class of entities, but with a whole class of classes, namely, with all classes having internal relations of some specified type. And by the *type* of a relation I mean its purely logical properties, such as are denoted by the words one–one, transitive, symmetrical, and so on. Thus for example we define the class of classes called progression by certain logical characteristics of the internal relations of terms of any class which is a progression, and we found that finite Arithmetic, in so far as it deals with numbers ... applies equally well to all progressions ... The so-called axioms of Geometry, for example, when Geometry is considered as a branch of pure mathematics, are merely the protasis in the hypotheticals which constitute the science ... My object in the present chapter is to execute the purely formal task imposed by these considerations, and to set forth the strict definitions of various spaces.

I leave aside for the moment all that which concerns the notion of a relational type[5] to focus on the parallel drawn between space and progression. It is clear that Russell applied here the manoeuvre we have just described: the geometrical postulates should be conceived as a nominal definition of a class of structures. A space, like a progression, is thus nothing else than a certain (second-order) propositional function – or a set of relational structures.

As I have shown in Chapter 1, Russell considered that real projective space was the most fundamental geometrical notion, and that Pieri's characterization was the right one. In a very consistent way, Russell, in chapter 49, rewrote Pieri's axiomatic in order to bring forward the 'logical' nature of the supposed geometrical indefinables. Thus, a projective line is characterized (see PoM, p. 430) as a certain relation which holds between any two elements of the set; it is symmetrical, connected (if a and b belongs to the field of R, then either aRb or bRa), and aliorelative (no term has the relation with itself); and it has further properties that are listed in Pieri's axioms. The term 'line' is then replaced by a variable satisfying certain axioms, exactly as 'successor', in Russell's logicization of Peano's arithmetic, was replaced by a variable fulfilling certain formal properties. Russell next introduces the incidence relations between points and lines in terms of the membership relation between the fields of relations that projective lines are. The relation of harmonicity is then defined by the quadrilateral construction, and all the other Pieri postulates are reformulated in this 'relational' way (see PoM, p. 431). Through this process, the concept of projective space is nominally defined, and in this definition only logical constants and variables occur. Note that the concept of a projective space is, like the notion of a progression, a second-order concept. It applies to a class of relational structures and not to a particular one. However, the insight that projective space is at bottom an incidence structure is not eliminated. On the contrary, what is regarded as essential to projective space is not the nature of its constituent elements, but the incidence relations between its 'points', its 'lines' and its 'planes'.

Now, projective space is not the only type of space logically defined by Russell. In PoM chapter 49, he shows in addition how to logicize Peano's vectorial axiomatic of three-dimensional Euclidean geometry (§ 414), and even the more exotic theory of Clifford's space of two dimensions (§ 415). In fact, as we have seen, any axiomatic characterization can be turned into a logical nominal definition of a class of structures. The elimination of the geometrical (and more generally, non-logical) indefinables is thus no more than a formal trick, which can be mechanically applied to any axiomatic system. And as Coffa was prone to emphasize, this causes a serious problem: if every axiomatization leads to a logical definition, why not extend the process of logicization to any axiomatized science? Nothing seems to rule out the possibility of axiomatizing geography – why not then consider geography as a logical science? More to the point: we have seen that Russell held that projective space is a logical concept, yet he regarded metrical space as an empirical notion. He knew that metrical geometry could be axiomatized. Why did he not turn the axiomatic characterization of Euclidean space into a logical definition?

At this stage, one might be tempted to refer to the second element distinguished by Russell in PoM § 474. To show that a given theory belongs to logic, it would be required not only to axiomatize it, but also to prove from logic that the theory in question has a model. The need to prove existence-theorems would act as an additional constraint allowing us to distinguish between the logical and the non-logical axiomatized theory. As we will see, however, this exit does not work.

3.2   Russell's if-thenism (II): existence-theorems

There are two ways of proving that a definition is not empty: one can either show that an empirical entity has all the required properties, or one can prove, from logical premises alone, that there is a model satisfying all the conditions. Let me quote again the introduction to Whitehead (1906b, p. 2), where this distinction is made:

> The Existence-theorem for a set of axioms is the proposition that there are entities so inter-related, that the axioms become true propositions, when the points are determined to be these entities and the relations between points to be these inter-relations. An Existence-theorem may be deduced from purely logical premises; it is then a theorem of Pure Mathematics; or it may be believed as an induction from experience; it is then a theorem of Physical Science.

The distinction between 'logical' and 'empirical' existence-theorems seems to provide an easy answer to the question I raised at the end of section 3.1: the logical theories would be the theories which can be axiomatized, and for which a logical existence-theorem is available. A formalized geographical theory would not be a logical science, because, even if it could be axiomatized, it will only have, at best, an empirical model.

Before explaining why this answer does not work, let me list the logical existence-theorems that Russell used in PoM. He enumerated these in the final part of § 474 (pp. 497–8):

> The existence of zero is derived from the fact that the null-class is a member of it; the existence of 1 from the fact that zero is a unit-class ... Hence, from the fact that, if n be a finite number, $n + 1$ is the number of numbers from 0 to n (both inclusive), the existence-theorem follows for all finite numbers ... From the definition of the rational numbers and of their order of magnitude follows the existence of η, the type of endless compact denumerable series; thence, from the segments of the series of rationals, the existence of the real numbers, and of θ, the type of continuous series ... From the existence of θ, by the definition of complex numbers ..., we prove the existence of the class of Euclidean spaces of any number of dimensions; thence ... we prove the existence of the class of projective spaces, and thence, by removing the points outside a closed quadric, we prove the existence of the class of non-Euclidean descriptive (hyperbolic) spaces ... Throughout this process, no entities are employed but such as are definable in terms of the fundamental logical constants. Thus the chain of definitions and existence-theorems is complete, and the purely logical nature of mathematics is established throughout.

One can distinguish three stages in this construction. First, Russell showed that there is a class for any finite cardinality. This amounts to saying that the infinite axiom is a logical theorem. As we know, in the framework of the ramified type theory, this is no longer true. A major consequence of the paradox is to compel Russell to step back on this crucial issue (see Boolos 1994).

But let us leave aside this point, and let us grant Russell that one has an existence-theorem for any finite cardinal. One can then move on to the next step: proving the existence-theorems for the different number systems, the rational numbers ℚ, the real numbers ℝ and the complex numbers ℂ. This is done in the standard way: from ℕ, using the set-theoretical resources then available (Dedekind's cut construction), one can develop arithmetical models for the different types of numbers. From this stage on, one reaches the third step: providing logical existence-theorems to the main geometrical concepts. The complex numbers give us a model of Euclidean space. One can then use Pasch's work to define projective space and construct an arithmetical model of the projective plane.[6] Once this stage is reached, it is very easy to develop arithmetical structures corresponding to the various metrical spaces by specifying different Absolutes. So Russell is right when he says that 'existence-theorems of mathematics are almost all obtained from arithmetic': the proofs that the mathematical spaces examined in Part VI are not empty are based on the fact that one can construct a numerical model of each of them. And as he believed (at the time) that one can derive from logical principles the concept of an ω-sequence that is not empty, all the existence-theorems listed above are purely logical.

Let me now come back to the main problem. The suggestion was to use the distinction between logical and empirical existence-theorems as an explanation for the fact that projective space is considered as a logical concept while metrical space is not (more generally as a means to demarcate the logical from the empirical sciences). The results listed above show that the purported solution is untenable. Logic in PoM is too strong: it supplies existence-theorems for projective space, but also for the classical metrical spaces as well. One then cannot hold that a theory is empirical if it does not have any logical model: metrical geometries have arithmetical models, but metrical geometry was nevertheless regarded as an empirical science. There is an even more general knockout argument against this proposal. By Gödel's completeness theorem, any first-order theory which is consistent has a set-theoretical model, and would then be considered as 'logical' by Russell, according to our purported standard – which is absurd. Taking into account the fact that Russell's logic is not first order makes things worse: owing to the incompleteness of second-order logic, the only way to show that a second-order theory is consistent is to exhibit a set-theoretical model. The lesson to draw from this analysis is that to prove that an axiomatic system has a 'logical existence-theorem' does not say anything about the empirical or logical character of the theory considered.

Let us take stock of what we have said so far. In section 3.1, I showed that any axiomatized theory can be regarded as a logical definition of a certain concept, and then that one cannot exclude geography, for instance, from the sphere of the pure sciences. In this section, I have shown that the attempt to base the demarcation between the logical and the empirical sciences upon the distinction between logical and empirical existence-theorems is a failure. Empirical axiomatized theories like geography, if they are not inconsistent, have 'arithmetical' models too. The question whether a given branch of knowledge belongs to the sphere of logic is not settled by the reference to the distinction between logical and empirical existence-theorems. We are thus led back to exactly the same situation we were in at the beginning of this chapter: we have not succeeded in extracting from Russell's texts any clear criteria allowing us to mark off logical from empirical knowledge.

Before moving on to a different approach, let me make three remarks. First, as I mentioned above, there is an evolution to the existence-theorems, from PoM to PM. In the ramified type theory, it is not possible to show that there is a class of any finite cardinality – a special axiom (the infinity axiom) is required to obtain this result. Much has been said about this change.[7] But let me emphasize that Russell's evolution does not affect my point. I'm seeking formal criteria that differentiate a logical theory, like arithmetic, from an obviously empirical one, like geography. The arguments I have used to show that one does not find any such criterion in PoM apply equally well in PM. Let us first assume that the infinity axiom is a logical proposition – then, an axiomatized geography (unless it is inconsistent) would have a logical model and, in this respect, it would be on an equal footing with arithmetic. Let us now assume that the infinity axiom is an empirical truth – then, an axiomatized geography would no longer have a logical model, but it would still be in the same position as arithmetic, which also no longer has any logical model.[8] In both cases then, the status of a purely mathematical science as arithmetic is aligned with one of the purely empirical sciences as geography. The distinction between logical and empirical knowledge cannot be based, neither in PM nor in PoM, on the difference between logical and empirical existence-theorems.

My second remark turns on the distinction Russell draws between definition and existence-theorem. Arithmetical models allow us to prove that the mathematical concepts we are interested in are not empty. But it would be a mistake to identify the mathematical concepts themselves with the arithmetical structures which exemplify them. For instance, a projective space is not a numerical manifold. As Russell noted, the existence-theorems are 'all obtained from arithmetic'; this does not mean, however, that the concepts whose existence is proved belong to arithmetic. As we have seen, Russell was wholly opposed to the Cartesian standpoint. This is why he promoted the view that space was at bottom an incidence structure: what is essential to a space is not the nature of its elements (they can be n-tuple of numbers, but many other thing as well), but the fact that these elements are linked together by some incidence relations. Scholars have not always been clear on this point and seem to have inferred from PoM § 474 the conclusion that geometrical space was for Russell a kind of numerical manifold.[9] It is true that, in the proof of the geometrical existence-theorems, analytic geometry played a central role. However, one should be cautious in differentiating the process of definition from the proof of an existence-theorem.[10] This distinction gives us the means to maintain that projective space should not be identified with a numerical manifold, while acknowledging that the only way to prove the required existence-theorem is to refer to the arithmetical models.

My third remark concerns the definition of finite cardinals. As I have already mentioned in the Introduction, this case is a special one. What we have just said about definition and existence-theorem helps us to see why. As is well known, Russell based his definition of cardinal numbers on what he calls the 'principle of abstraction': let R be an equivalence relation, then xRy if and only if there is an entity a and a many–one relation S such that xSa & ySa (PoM, p. 220). As Landini notes (2006, p. 231), 'the principle says that for any exemplified equivalence relation R there is a function $[S]$ which chooses a distinct entity $[a]$ to represent each equivalence grouping'. The problem is that this principle, which is considered as a logical theorem by Russell, does not determine which among the many possible representatives one should regard as '*the* number of the classes in question' (PoM, p. 114). All the different representatives a to which x (similar to u) has a certain many–one relation S is a good candidate for being the reference of the phrase 'the number of x'.[11] Russell (1901a) was very clear on this point when he introduced the class of all the couples $\langle a, S \rangle$ such that if x is similar to y, xSa & ySa.[12] Now in PoM, Russell chooses one couple among all the possible ones, since he held that the entity a ('the number of u') is the class of all the classes similar to u (and accordingly, that the relation S is the relation of set membership). As Landini explains (1998, pp. 27–8), Russell's decision was based on the fact that the reference to the class of all the classes similar to u is needed in the proof of the principle of abstraction.[13] Now, the principle of abstraction is an existence claim (R being an equivalence relation, xRy iff *there is* an entity a and a function S such that xSa & ySa). And to say that Russell's choice of a particular couple $\langle a, S \rangle$ is guided by the need to prove the principle of abstraction amounts then to saying that his choice came from the wish to prove the existence of at least one such couple. In the case of the cardinals, the definition is intimately connected with the existence proof. This marks out this case from all the other ones. As we have just seen, in PoM, the definitional stage is usually independent of the process of proving the required existence-theorem. There is one place, however, where this rule is not followed: the definition of finite cardinals.[14] In this case, the instance that allows us to show that there is at least one entity satisfying the relevant conditions is taken as the reference of the defined concept. But once again this is an exception, not the rule – a fact that the focus on the definition of finite cardinals in the literature leads us to forget.

3.3 Topic specificities and psychologism

From the inquiry conducted so far, we are led to conclude that the two criteria which are given (a logical theory should be axiomatized and should have a logical model) in § 494 are not sufficient to demarcate the logical from the empirical sciences. These are much too weak to avoid the kind of trivialization Coffa pointed out. As Russell and the Russellian scholars do not allude to any other formal constraints, one should conclude that some non-formal features have to be taken into account in the drawing of the boundary.

This question about the divide between logical and empirical science is a part of a more general sort of issue. The two previous chapters have shown that, to distinguish between the different alternative analyses he considered, Russell heavily relied on some non-formal considerations. Pasch's theory of projective space is formally perfect – but Russell still preferred Von Staudt's characterization. The 'arithmetical' conceptions of projective or metrical spaces are formally perfect – Russell nevertheless rejected the Cartesian approach. Metrical space can be defined as a particularization of the real projective space – but Russell did not take this path. In all these cases, Russell recognized that the analyses he rejected were formally correct. His decisions to reject them were then not based on formal reasons, but always on the considerations of some local and topic-specific features of the targeted theories. In a sense, this is good news: it shows that a Russellian analysis is not a mechanical process, blind to the particular properties of the body of knowledge it is applied to – that is a complicated development, which has to integrate many heterogeneous ingredients. But this recognition also raises an immediate difficulty, since it seems to clash with a fundamental tenet of Russell's philosophy: his anti-psychologism.

In PoM, Russell draws a sharp distinction between the conceptual content of a sentence and the way our mind apprehends this content. For instance, he starts his analysis of counting by stressing the need to distinguish between the logical and the psychological aspects of the question (p. 114):[15]

> What is meant by counting? To this question we usually get only some irrelevant psychological answer, as, that counting consists in successive acts of attention ... Counting has, in fact, a good meaning, which is not psychological.

Russell makes it clear that the logical analysis does not aim at preserving all the ideas that the usual mathematicians associated with a concept. Indeed, some of these ideas are not part of the genuine conceptual content, but are related to it merely by some kinds of psychological associations.[16] The need to differentiate the logic from the psychology belonged to the basic toolkit Russell drew on when he had to attack some rival position – thus (PoM, p. 4):[17] 'Philosophy asks of Mathematics: What does it mean? Mathematics in the past was unable to answer, and Philosophy answered by introducing the totally irrelevant notion of mind'. But Russell also used this distinction as a defensive weapon, allowing him to rebut certain objections faced by his programme. The manoeuvre was always the same: he criticized his opponents for having been unable to cut off the logical content from its psychological accompaniments.

A telling example of this defensive strategy is furnished by the answer Russell (1905b) gave to the criticisms Pierre Boutroux[18] addressed to his theory of real analysis. As we will see, his dispute with Russell should be seen as part of the long quarrel between Russell and Poincaré. A mathematician himself, Boutroux held the chair of history of sciences at the Collège de France from 1920 to 1922. He wrote an important book on mathematical analysis, *Les principes de l'analyse mathématique. Exposé historique et critique* (1914–19), in which he attempted to combine a theoretical with a historical perspective. His 1905 paper is a first step in this direction. Its purpose was to defend an 'intermediate' view of function, according to which function should neither be reduced to the concept of algebraic equation (as Kronecker would have it), nor be defined as a kind of relation (as Russell claimed). For Boutroux, the notion of a function that one found in mathematics was much more refined than what Russell seemed to believe. In particular, Boutroux opposed Russell's characterization of real analysis as a theory of order. Thus, he wrote (1905, pp. 628–9):

> Contemporary logicians succeed to construct, with the help of ordinal notions, the theory of irrational numbers ... From this, Russell and Couturat seem to conclude that the whole of real analysis boils down to the science of order. This conclusion, it seems to me, is not valid ... The study of the continuum is only the introduction to real analysis. There is no doubt that, in real analysis, there are variables which take their values within continuous series; but these variables have a role to play only in so far that some questions of an entirely new kind are raised about them. The philosopher who endeavors to deal with the notions which lie at the basis of real analysis should focus on these questions – not on the continuum.

As an example, Boutroux mentioned the theory of differential equations. This doctrine played a prominent role in the development of the differential calculus – yet it seemed that the theory of differential equations has nothing to do with order.[19]

Boutroux's position was subtle. The mathematician did not reject the possibility of reducing real analysis to the theory of ordinal relations. What he pointed out was that Russell's exclusive focus on order could not account for the kinds of function mathematicians actually worked on, and could also not explain the reasons why scientists have studied certain functions rather than others. Poincaré, in 1908, espoused exactly the position.[20] He did not reject the fact that a mathematical concept can be ultimately decomposed in logical constituents, but emphasized that logic itself (the theory of order, in the case in point) did not provide the reason why the different logical ingredients are brought together in the way they are in mathematics. Let me quote this passage (1908, p. 126):

> Our body is composed of cells, and the cells of atoms, but are these cells and atoms the whole reality of the human body? Is not the manner in which these cells are adjusted, from which results the unity of the individual, also a reality, and of much greater interest? Would a naturalist imagine that he had an adequate knowledge of the elephant if he had never studied the animal except through a microscope? It is the same in mathematics. When the logician shall have resolved each demonstration into a host of elementary operations, all of them correct, he will not yet possess the whole reality; this I do not know what which constitutes the unity of the demonstration will escape him completely. In the edifices built up by our masters, of what use to admire the mason's work in the edifices erected by great architects, if we cannot understand the general plan of the master? Now pure logic cannot give this view of the whole; it is to intuition we must look for it.

Russell replied to Boutroux's and Poincaré's attack in the following way (1905b, p. 270):[21]

> The confusion I believe I have discovered between the motives of study and the objects studied appears in many of the arguments directed against the reduction of mathematics to logic ... The point is that no one pretended that logic alone could tell us what problems we should study, and that the choice of problems to study is, everyone will agree, a matter of individual judgment. But this has nothing to do with the question of knowing if the solution of mathematical problems can be obtained in terms of logic ... What Mr Boutroux says is true, applied to the psychology of the mathematician; but the psychology of the mathematician is not mathematics.

Russell's answer is quite unsurprising. In a more up-to-date terminology, it consists in bringing forward the distinction between 'context of discovery' and 'context of justification'. Boutroux criticized Russell for not having accounted for certain important features of the mathematics he pretended to have reduced. Russell retorts that those features do not really belong to the mathematical content, but only to the psychology of the mathematicians. It is true, said Russell, that logic does not teach us what functions and what problems we should study; it is for instance true that the way real analysis is developed in PoM does not explain why mathematicians have developed a theory of differential equations; but all this does not matter, because the logicists do not aim at accounting for these facts. Indeed, these features concern the way we apprehend the mathematical content – not the content itself.[22]

Russell's rebuttal of Boutroux relies then on the distinction between mathematics itself and the psychology of mathematics. Now, is it always that easy to draw a line of separation? In a sense, if-thenism provided a positive answer to this question. If the theorems of a given theory T can all be derived from a certain number of axioms, one can then consider that the content of T is already given in the axioms – even if the 'questions' mathematicians ask in T have nothing to do with the concepts occurring in the postulates used to axiomatize T. For instance, assuming that real analysis is completely axiomatized by the postulates lying at the basis of the theory of continuity, one could claim that the 'content' of the theory of differential equations is contained in the theory of continuity because the axioms needed to characterize the latter are sufficient for characterizing the former. Of course, there are many directions in which one could develop an axiomatic system, and these different ways are not displayed in the list of axioms. But all this becomes irrelevant as soon as one focuses on the content of the theory (i.e. on what is given in the axioms), and not on the manner we apprehend and develop this content. Thus, even if it is true that, as Boutroux emphasized, the ways of combining the primitive notions in order to form the high-level concepts of real analysis are not given in the basic concepts themselves, and even if the ways of combining the axioms to prove the important mathematical theorems are not explained in the primitive propositions themselves, it remains that the 'content' of real analysis is contained in the theory of continuity. If-thenism would then give us the means for distinguishing once and for all what belongs to the mathematical content from what pertains to the psychology of mathematics.[23]

Now, the close examination of PoM VI we have carried out in the first two chapters show that if-thenism is a much too formal and too mechanical conception of Russell's analytical practices – and that Russell knew how difficult it was to separate the logical grain from the psychological bran. Let me recall the facts.

First, we have seen that Russell presented two constructions of projective geometry, one maintaining Pieri's approach, the other following Pasch's view. From a mathematical point of view, the two perspectives are equivalent, in the sense that they define exactly the same structure. It seems thus that Pasch's approach, based on order relation, would better suit Russell's project. Indeed, PoM IV is entirely dedicated to the concept of order and it would have been very elegant to ground geometry on this basis alone. But, as we know, Russell chose to follow Pieri instead, and to characterize projective space as an incidence structure, arguing that only this definition enabled him to secure the independence of projective geometry. Owing to what Russell said in his reply to Boutroux, this poses a problem. Why attach so much importance to a feature (the independence of projective geometry) which seems to concern more the way the theory is presented than its 'content'? In Chapter 1, I laid stress on Russell's early involvement in Von Staudt's tradition. But this sort of biographical consideration is typically something one can regard as pertaining to Russell's own psychology of mathematics – it is typically something that should not play a role in the logical reconstruction of geometry. Why then did Russell put such a weight on a feature (his own view of the organization of the mathematical field) that he elsewhere (in his answer to Boutroux) attempted to downplay?

Second, after having developed the theory of the real line in PoM V, Russell could have developed an algebraic Cartesian approach to projective space – and more generally, of the various kinds of space. We have seen above that he alluded to this possibility in PoM § 474. One of the central theses in PoM VI is however that space must not be reduced to a numerical manifold. This interpretation could be challenged by exactly the same argument Russell had raised against Boutroux. Since analytical geometry provides us with a model of the various spaces, why not view geometry as just an extension of real analysis? Of course, some features of the geometrical practices will be lost in the operation – but who cares? The content of the various geometries (the truth value of all the theorems) would be preserved and, taking up again his distinction between psychology and logic, Russell could have written off what is eliminated in the translation as a mere psychological residue. Had he taken this road, the structure PoM VI (and then of PoM) would have been considerably simplified. The puzzle is that he did not take this path. Why did he not do so? He apparently thought that geometry should be considered as an independent branch of knowledge, irreducible to arithmetic. But what is the logical ground for this demand? Why did he not apply to this view the kind of argument he was so prompt to level against Boutroux?

Third, Russell distinguished two conceptions of metrical geometry in PoM: the first one construed metrical geometry as a marginal corner of projective geometry, while the other viewed the theory as an empirical science dealing with the properties of an empirical concept, the magnitude of divisibility. Unlike what happened in the two previous cases, Russell did not contrast two ways of reducing a given branch of mathematics to logic, but opposed a logical conception of metrical geometry to a view which conferred to metrical geometry the status of an empirical science. One might have expected that Russell would give priority to the first option. After all, according to Occam's razor, one should give advantage to a solution which does not require the introduction of any new indefinable – furthermore, as a logicist, Russell would certainly not have complained against this extension of the logical sphere. And again, the distinction between logic and psychology would have provided him with an easy way to defend his decision. He could have indeed maintained that men discovered the geometrical truths through the study of the properties of physical space – while arguing that, from a logical point of view, the concept of distance remained ultimately a projective notion. As we have seen, however, he chose another path. Since the projective derivation of distance did not account for the central place metrical geometry has in our knowledge, he decided to construe it as primarily an empirical science. Once again, the same problem resurfaces: why attach so much importance to a fact which seems to concern more the organization of human knowledge than the content of a theory? Drawing his inspiration from his answer to Boutroux, Russell could have objected to himself that the fact that metrical geometry played a central role in history should not be taken into account in a logical analysis.

These three cases illustrate the same difficulty. Russell's answer to Boutroux shows how prompt he was to smoke out in any non-formal consideration an irrelevant remark concerning the psychology of the mathematicians. The problem is that he himself took into account, in his logical analyses, what appears at first as some non-formal features. Why did he not address to himself the same argument he had launched against Boutroux?

The problem does not merely come from the fact that Russell drew a distinction between the logical content and the psychological scoria accidentally linked with it. It is also that, in his anti-psychologist attacks, he proceeds as if it is always obvious that the logical content can be kept separate from the psychological accompaniments. This is not true. And he had no general recipe for settling this issue (in particular, he was not an if-thenist). Here lies the genuine force of Boutroux's and Poincaré's position: contrary to what Russell (1905b) suggested, the task of delineating what is logically 'relevant', in a given part of mathematical knowledge, is not trivial. In fact, as is shown in the three developments above, Russell's practices of analysis wonderfully illustrated this point. The major part of his investigation in PoM VI was devoted to dealing with questions such as: what makes projective geometry specific? How can we view the relation between an arithmetical model and a geometrical space? What is the real topic of metrical geometry? The sole fact that he addressed at length these issues shows that he did not consider the distinction between logical content and psychological artefact to be an immediately given datum.

It remains however that Russell does not seem to have any global solution to the problem of making the demarcation between what does and what does not belong to the mathematical content. He supplied answers in specific cases. One can thus retrace the reasons as to why he favoured Pieri rather than Pasch, why he did not endorse the Cartesian approach to geometry, and why he assumed that metrical geometry was, at bottom, empirical. But he did not provide any general argument. Hence, his analyses remained exposed to the same kind of attack that he launched against Boutroux – they could be criticized for being the result of a confusion between logic and psychology.

3.4 The types of relations and the architecture of mathematics

The question we are facing is the following one: how can we reconcile Russell's anti-psychologism with his analytical practices? To meet this challenge, we should find a way of establishing that the topic-specific features he did take into account in his analyses were not purely psychological. At first, the task seems beyond reach. Indeed, at the beginning of PoM, Russell defined mathematics as 'the class of all propositions of the form "p implies q", where p and q are propositions containing one or more variables, the same in the two propositions, and neither p nor q contains any constants except logical constants' (p. 3). Complete generality was thus acknowledged as the characteristic feature of mathematics – and this fact seems to ruin any attempt to make room for a notion of topic specificity, since it seems to imply that a mathematical proposition has no particular subject matter. Now, this argument is not as convincing as it might seem at first. Indeed, as I will now show, Russell did not reject the idea that mathematics was divided up into different branches, and that each mathematical field had its own subject matter. What he claimed was that the various purely mathematical topics should not be conceived as non-logical constants, but as logical relational types.

It is in PoM § 8 that Russell introduces the notion of relational type (pp. 7–8):

> The process of transforming constants in a proposition into variables leads to what is called generalization, and gives us, as it were, the formal essence of a proposition. Mathematics is interested exclusively in *types* of propositions ... Whenever two sets of terms have mutual relations of the same type, the same form of deduction will apply to both. For example, the mutual relations of points in a Euclidean plane are of the same type as those of the complex numbers; hence plane geometry, considered as a branch of pure mathematics, ought not to decide whether its variables are points or complex numbers or some other set of entities having the same type of mutual relations. Speaking generally, we ought to deal in every branch of mathematics, with any class of entities whose mutual relations are of specified type; thus the class, as well as the particular relations considered, becomes a variable, and the only true constants are the types of relations and what they involve. Now a *type* of relation is to mean, in this discussion, a class of relations characterized by the above formal identity of the deductions possible in regard to the various members of the class; and hence a type of

relations, as will appear more fully hereafter, if not already evident, is always a class definable in terms of logical constants.

A type does not here designate the range of significance of a variable, as will be the case later on. A relational type, in the sense of PoM, is a class of relations characterized 'by some property definable in terms of logical constants alone' – thus, for example, 'one–one, many–one, transitive, symmetrical, are instances of types of relations' (ibid.). Furthermore, Russell suggests that the relational type provides the mathematical theory with its 'topic': what we deal with 'in every branch of mathematics' is a 'class of entities whose mutual relations are of specified type'.

The content of § 8 does not contradict the thesis brought forward in § 1. Mathematics is still characterized by complete generality. Yet, the introduction of types changes the deal, since it shows that the appeal to complete generality does not abolish the differences within mathematics. Thus, in the completely general conditional proposition, the antecedent characterizes a certain relational type. For instance, any proposition of plane Euclidean geometry has the form $T \Rightarrow p$, where T (the planar Euclidean axioms) defines the relational type of the Euclidean planes. Thus, a proposition of Euclidean geometry is both a completely general proposition and a proposition which deals with a particular topic, the relational type of Euclidean geometry. That the distinction between mathematical topics should not be based on the difference between particular non-logical constants does not mean that there are no such differences. This is the crucial point: in pure mathematics, 'the only true constants are the types of relations and what they involve'. A type is a constant to the extent that it gives a mathematical discipline its particular subject matter. At the same time, though, it is completely general and does not contain any non-logical terms. That plane geometry, considered as a branch of pure mathematics, 'ought not to decide whether its variables are points or complex numbers' does not mean that Euclidian geometry has no subject matter. It only means that Euclidean space is not a non-logical constant (an object of intuition), but the relational pattern that all the different models of Euclidean geometry (be they numerical or empirical) have in common. In many respects, this example of the Euclidean plane is badly chosen, since Euclidean geometry (more generally metrical geometry) is considered by Russell to be an empirical science.[24] But the general point is clear: to define pure mathematics as the science of generality does not commit us to getting rid of the notion of mathematical topic – it only requires characterizing it as a logical relational form.

This fine-grained description of Russell's view opens the way to a resolution of our initial problem. The difficulty was that every attempt to take into account certain non-formal features of a given part of mathematics could be dismissed as resulting from a confusion between mathematics and the psychology of mathematicians. But the notion of a relational type explains why certain non-formal properties can nevertheless be viewed as logical features. Let me explain how by taking the examples of Russell's analysis of projective space. As we have seen in Chapter 1, Russell's discussion was directed by the idea that one must preserve the purity of projective geometry as an independent discipline. This wish was linked to his early commitment to Von Staudt's tradition of pure synthetic geometry, and, understood in this way, his demand could easily be dismissed by an anti-psychologist move: a logical reconstruction of mathematics should have nothing to do with the subjective preference of those who undertake it. Now, to justify his analysis, Russell did not content himself with drawing attention to Von Staudt's and Pieri's works. He also referred to the fact that his new logic furnished the means to characterize incidence as a relational type. What he brought forwards in PoM VI is that the independence of the projective geometry as a discipline is reflected at the logical level in the distinction between relational forms: incidence relation, which is the 'topic' of projective geometry, is different from ordinal relation, which is the 'topic' of real analysis. The need to endorse Von Staudt's pure approach was thus grounded in PoM on a logical distinction between types of relation. In this regard, Russell's reasoning is very different from Boutroux's. Boutroux contented himself with referring to certain parts of existing mathematics (the theory of differential equations) without attempting to explain why this mathematical field should be taken into account in the logical reconstruction of real analysis. This absence left room for Russell's anti-psychological counter-attack. In PoM VI, Russell did not only point out some aspects of what mathematicians did, but he showed also how these researches could find a natural place within his relational framework. The anti-psychological attack Russell levelled against Boutroux cannot then be turned against his own analysis of projective geometry: in 1903, projective space is a particular relational type.

In Gandon (2008), I extended this reading to the whole of PoM. I argued there that each part of PoM (except Part I) is correlated to a certain relational type. Part II, dedicated to cardinal arithmetic, is a study of the properties of equivalence relations; Parts IV and V, bearing on real analysis, is a general theory of ordinal relation. We have just seen that PoM VI is a theory of a new type of relations, the incidence relations. There remain two cases to examine: Part III about quantity and Part VII about mechanics. Concerning the latter, Russell asserted that a material object, the fundamental concept of mechanics, is a 'three-cornered relation' between an entity, a place and an instant. He seemed then to tie a link between mechanics and the idea of ternary relations.[25] Concerning Part III, one will see in Chapter 4 that, as far as mathematics is concerned, the notion of quantity has some essential connection with the concept of relative product (i.e. with the notion of composition of relations). One can then substantiate the idea presented in PoM § 8 by proposing a table of correspondence, assigning to each part of PoM its own relational type (see Table 3.1).

Table 3.1   Correspondence between the parts of PoM and the relational types

Cardinal arithmetic PoM II	Theory of quantity PoM III	Mathematical analysis PoM IV–V	Geometry PoM VI	Mechanics PoM VII
Equivalence relations	Relative product	Order relations	Incidence relations	Three-termed relations

Of course, the associations made in Table 3.1 must be taken with a grain of salt. If one can defend the idea that, in PoM, real analysis is concerned with ordinal relations and geometry with incidence relations, it seems more complicated to sustain the claim that cardinal arithmetic is merely about equivalence relations, that theory of quantity is only about relative products and (above all) that mechanics is just the theory of certain three-termed relations. Yet, it remains that even in the case of mechanics (admittedly the most disputable one) Russell attached some importance to the existence of a connection between the branches of mathematical knowledge examined in the different parts of PoM and some particular types of relation.

Russell's logicism is no doubt a reductionism: in PoM, all the mathematical knowledge was conceived as an extension of the new logical theory of relations. But what is noteworthy is that the Russellian reduction aimed at preserving the organization of the mathematical field. In other words, logic was not only a means to derive all the mathematical truths – it was also an instrument for representing how the mathematical sciences were divided up. In this respect, the logicist agenda was very different from the arithmetization programme. The arithmetizers aimed at reducing the whole of mathematics to one of its branches, arithmetic – for them, and for most scholars today, reducing mathematics to logic would amount to reducing arithmetic to logic. In this view, arithmetic would be the only channel through which logic and mathematics interacted, and the theory of relations would not express the differences between the various existing mathematical disciplines. One could even say that there is a straightforward opposition between logicism and arithmetization. Indeed, to reduce mathematics to logic is a way to restore, against the threat represented by the reduction of all mathematics to arithmetic, the specificities of the various disciplines. In PoM, the independence of the main mathematical fields (arithmetic, real analysis, geometry) is granted a logical status. And in this perspective, the erasing of the differences between mathematical topics becomes a logical mistake. Claiming for instance that space is no more than a numerical manifold amounts to mixing up two different logical forms, i.e. incidence and ordinal relation. In Russell's system, the way the logical theory is used in Part II (devoted to cardinal arithmetic) differed from the way it is used in Parts IV and V, dedicated to real analysis – which itself differed again from the way it is developed in PoM VI. The theory of relations, as it is developed in the various parts of the system, is each time specific, and the distinctions within pre-logicized mathematics (between real analysis and arithmetic, between geometry and real analysis, etc.) are, in this manner, internalized into the relational framework (i.e. reflected by the distinctions between relational types). To bring down the barriers between the disciplines amounts then to missing the way the logical content is structured.

As we have seen, Russell's logical translation was blamed by Poincaré and Boutroux on his ignoring certain important aspects of mathematical practices. This critical line, according to which the logical rewriting betrays the mathematical practices, was radicalized by the second Wittgenstein. Indeed, Wittgenstein emphasized that the multiplicity of the techniques of proof, of mathematical notations, etc., was an essential ingredient of mathematics (see for instance 1956, p. 84): 'I should like to say: mathematics is a MULITICOLOURED *mixture* of techniques of proof'. From this perspective, Russell's logicism was viewed as an effort aimed at systematically hiding the differences between mathematical language games (ibid., p. 89): 'If someone tries to shew that mathematics is not logic, what is he trying to shew? He is surely trying to say something like: If tables, chairs, cupboards, etc. are swathed in enough paper, certainly they will look spherical in the end.' Russell could have countered this attack by using the same kind of argument that he used to brush aside Boutroux's reasoning: the purported mathematical diversity is only apparent; it belongs to the psychology of the mathematicians, not to the mathematical content. But this uninteresting answer would have been contradicted by Russell's own analytical practice. What I have suggested here is that Russell had at his disposal a more promising line of defence: far from aiming at making mathematics uniform, the logical reduction aimed at reflecting in the logical system the fine-grained differences among the mathematical field. Recall that Russell went very far in this direction: projective geometry could have been seen as an extension of arithmetic, it could also have been grounded on the notion of order – but he decided to grant it its own relational niche, in characterizing it as a pure theory of incidence. To suggest, as Wittgenstein did, that Russell endeavoured in his work to swath 'tables, chairs, cupboards, etc.' in enough paper, so that they all 'will look spherical in the end' shows thus a complete misunderstanding of what Russell's logicism was. The logical theory of relations was not a steamroller which flattened all the variety of the existing mathematical practices. Far from considering the attempts to save the distinctions between mathematical disciplines as a consequence of a confusion between psychology and logic, Russell used the new logic to represent these pre-logicized differences. His reduction did not only aim at developing the mathematical material from logic, but also at preserving the mould into which this material was cast. To logicize mathematics is not to uniformize mathematics.

At this stage, a natural question arises: Russell gave a logical status to the traditional distinctions between mathematical branches – but how did he view the status of the science that logic itself began to be at that time? In Table 3.1, each part, except the first, is corresponded with a relational type – but what about PoM I? To which relational form should we correlate the science of logic? Russell (1901b, p. 357) gave some clues about his way of conceiving logic as a branch of mathematics:

> A Symbolic Logic may be considered from three points of view: (1) as regards its inherent correctness, (2) as regards its suitability for application to actual examples of deductive reasoning, (3) as a convenient or inconvenient instrument in building up the Logical Calculus as an independent branch of Mathematics.

After showing that, in the first two respects, Peano's method (maintained in PoM) is a 'vast improvement on its predecessors', Russell comes to the third point (ibid., p. 358):

> Considering the Logical Calculus as an end in itself, we miss in Peano's system the complete reign of duality as between addition and multiplication, which forms one of the most attractive features of the subject ... It would be easy, by a few modifications, to introduce as much as duality as is true, but where formal logic is a mere preliminary to other branches of mathematics, the course followed in the *Formulaire* seems on the whole the most desirable.

In Peano, logic does not appear as an independent branch of science. This is not, for him, a drawback, however, since formal logic must be considered as 'a preliminary to other branches of mathematics', and not as an independent mathematical field ('as an end in itself'). For Russell, the theory of relations can thus not be put on a par with the other mathematical disciplines: in order that logic becomes the framework within which the other branches could be developed, one must renounce seeing logic as an independent discipline. Logic is not regarded in PoM as a mathematical branch. It is used by Russell as a mirror destined to reflect the whole of the mathematical architecture.

Let me come back to our initial question: how can we explain why Russell gave so much importance to some non-formal considerations which could easily be dismissed as pertaining to the psychology of mathematicians? In PoM § 8, Russell connected the specificities of a given mathematical branch to the logical properties of a particular relational type. In so doing, he succeeded in giving to the non-formal considerations he relied on to support his analysis a logical status. The idea that projective geometry (for instance) is an independent branch of knowledge was not justified by a reference to the distinctive character of a certain mathematical tradition – it was grounded in PoM on the definition of the projective space as an incidence structure. This notion of type of relation was then crucial since it allows Russell to escape the threat of psychologism. But a difficulty remained, which directly stems from the way the concept of relational type was characterized. I will explain this in the next section.

3.5 The interaction between logic and mathematics: a persistent difficulty?

A relational type is a hybrid notion. It is defined by two different characters, which can exist independently of each other. It is first and foremost a purely logical notion: any class of relations definable in logical terms is susceptible to becoming a relational type. But on the other hand, the notion is connected to a certain state of the development of existing mathematics: relational types should correspond to the topic of a particular mathematical branch. These two properties are completely separable. There is no reason to think that a logically definable class of relations is always at the basis of a mathematical discipline; and conversely, there is no reason to think that all pre-logicized mathematical fields can be defined as the study of a particular relational logical form. To illustrate the idea, let me take some examples.

In PoM, a relation is said to be connected if it is such that any two elements of its field can be related by a certain finite number of steps (pp. 202, 239). For instance, a graph is said to be connected if there is a path which relates any two of its vertices. Russell used this concept when dealing with order, but he did not develop a general theory of connected relations. The reason he did not do so was presumably that this class of relations was in 1903 not linked to a particular mathematical discipline. In this regard, the situation with respect to connected relation was thus different from the situation with respect to ordinal and equivalence relations. Had PoM been written 50 years later the situation might have been different. One could indeed hold that connected relations play a fundamental role in graph theory.[26] Had he known graph theory, Russell could have been tempted to define it as the study of 'connected relations' and to dedicate an entire part of PoM to its presentation. But, in 1903, graph theory did not exist as an independent field of research, and Russell did not feel the need to study in depth the properties of connected relations. This example shows that not all logically definable classes of relations are relational types. It is the existing state of the mathematical sciences which determines which logically definable classes of relations, among the indefinite many ones, should be considered as genuine relational types.

But the converse is also true: some existing mathematical fields do not seem to fit into the relational framework. According to Boutroux, the theory of differential equations provides us with a good illustration. One does not see what could be the class of relations corresponding to this field of research. One knows Russell's answer: if the new logic does not accommodate the theory of differential equations, it is because this domain is too advanced to be regarded as one of the basic component parts of mathematics. But Boutroux's criticism could be interpreted as pointing out the fact that there is no consensus on the way mathematics should be divided. It was not absurd, at the beginning of the twentieth century, to hold that a theory of differential calculus which does not say anything about the theory of differential equations must be rejected. And it was not absurd either to support the analytic approach of geometrical space which PoM VI was wholly opposed to. Russell never explained why one should, in some cases, consider a certain mathematical discipline as an independent science (projective geometry), while, in some other cases, this autonomy should be regarded as a mere psychological artefact (the theory of differential equations). The reference to existing mathematics is far from solving this issue, since, most of the time, as Boutroux's objection shows, mathematicians disagree about the importance of a given topic, about the relevance of a particular distinction. The architectural data that pre-logicized mathematics supplies are too faint, too vague and sometimes even too contradictory to be used as reliable material.

The situation is then odd: to know whether a logically definable class of relations must be studied in depth in the system, one has to look at the actual state of mathematics and see if a mathematical discipline can be correlated to it; but to know whether a certain mathematical field could be regarded as a genuine independent science, one has to find in the logical repertoire a type which can correspond to it. There is thus a dangerous circularity in Russell's reasoning. A reference to existing mathematics is needed to determine whether a given relational form is a type; but the structure of pre-logicized mathematics could not be fixed independently of the relational framework. How can we break this circle? What comes first: the logical machinery or the actual state of mathematics? Russell seems to hold that the logical reduction is not a mere internalization of a content given outside, in the existing mathematics. At the same time, he does not derive the structure of pre-logicized mathematics from the logical scaffolding. How then did he succeed in adjusting the relational framework to the internal division of existing mathematics? How did he proceed to match the main mathematical disciplines to some relational forms?

This difficulty was obscured in section 3.4 above because I gave the impression there that the differences between relational types reflected the organization of existing mathematics. I thus implicitly assumed that the information flow went in PoM from the existing mathematics to the logical system: the theory of relations was presented as a passive receptacle aimed at representing a content given from outside, as 'a mirror destined to reflect the whole of the mathematical architecture'. But this is not true.

As I have just mentioned, there was no consensus in 1903 (and there is still no consensus today) on the way mathematics should be organized, and Russell relied on his relational framework to single out, among the possible ways of articulating pre-logicized mathematics, the one which suited him best. In PoM, logic was not only a passive receptacle – it was instrumental in carving up mathematics.

To show this, let me take the example of Russell's reception of Pieri's work. At first, one could consider that the PoM theory of projective space illustrates the fact that Russell trailed behind the advances in existing mathematics. Had Pieri not written his treatise (had Russell not read it), the structure of Russell's logical system would not have been the same.[27] One should how-ever emphasize that, apart from Russell, nearly no one at the time grasped the strength of Pieri's insight. Veblen, in 1906, published a very brief article (three pages), entitled 'The Square Root and the Relation of Order', whose explicit purpose was to 'infer order relations in terms of postulates about a square root'. Translating Pieri's construction in algebraic terms, Veblen succeeded in giving a definition of the real number field which did not introduce any reference to an ordinal relation. Now, as Sinaceur (1991) has shown, Veblen's (Pieri inspired) idea anticipated by nearly 20 years some key ingredients of Artin and Schreier's (1926) theory of the real field.[28] So, the insight that order can be defined in terms of incidence, the very insight that Russell brought forward in PoM, proved to be a very fruitful one. But if so, how can we explain that (with the notable exception of Veblen) no one at the time showed interest in this insight? What led Russell to under-stand what the contemporary working mathematicians overlooked? It is the importance he attached to relations which led him to ask the good question (how is it possible to ground projective geometry on incidence, instead of order?), and thus to understand the significance of Pieri's achieve-ment. This shows that the new logic of relation, far from being a passive receptacle, provided Russell with an interpretative grid which allowed him to read existing mathematics in a particular way, to point out neglected relevant distinctions and to find new ways of organizing the mathematical material. Had he not read Pieri, PoM VI would have been different – but had he read Pieri without having in mind his theory of relations, he would not have been able to grasp the strength of his approach. The case of projective geometry illustrates in a vivid way that the theory of relations played an active role in the structuration of the mathematical field. Russell did not start then with existing mathematics. On the contrary, his relationship to pre-logicized mathematics was informed by the new logic.

One should however not go too far in this direction and believe that Russell could derive the structure of pre-logicized mathematics from the theory of relations alone. In PoM, the logical framework did not govern the articulation of mathematics. A comparison between the role relational logic had in PoM and the role abstract algebra played in the structuralist revolution occurring in the mathematics of the 1920s could help us to understand this point. Spurred on by Emmy Noether's works on abstract algebra, the mathematicians of the interwar period completely reorganized the architecture of classical mathematics: the old traditional organization, grounded on the differences between arithmetic, geometry, differential calculus, etc., was completely supplanted by a new framework based on the distinctions between the various algebraic structures (groups, rings, fields, etc.) and on the systematic study of their properties and relationships.[29] Such a revolutionary move did not occur in PoM. The theory of relations was not developed for its own sake, but only in so far as it led to a capturing of the content of the already existing mathematical branches. Russell did not attempt to project the relational scaffolding onto existing mathematics; his aim was not to reshape the entire mathematical field (the Tables of Contents of PoM and PM are quite traditional); on the contrary, as I explained in section 3.4, logic was seen as a means of displaying the essence of the various traditional theories. Russell's logicism was a philosophical project, not a mathematical revolution: logic did not rule mathematics as the algebraic structures ruled mathematics in the 'structuralist' reform of the 1920s.[30]

In other words, Russell did not want to derive the organization of existing mathematics from logic, but he did not want either to import within the logical system the pre-logicized outlines of the existing mathematical disciplines. His effort was to strike a balance between the relational framework, on the one hand, and the pre-logicized mathematics, on the other. Neither of the two components of the relational type had a priority over the other: logic was not a passive receptacle whose sole role was to reflect distinctions present outside it – yet, the structure of pre-logicized mathematics could not be read off the relational framework. This leaves us with a problem: how can we balance, in particular cases, the importance of the two factors? In a relational type, the logical framework is supposed to be adjusted to a certain branch of mathematics. But how is this alignment between mathematics and logic supposed to be achieved? To reach such equilibrium, some tailoring works are necessary: one has to recast certain parts of mathematics by adjusting them to the logical resources; conversely, one has to develop the study of certain particular relational forms to the detriment of others in order to meet real mathematics.

Relational types are not data which one can rely on before the analysis takes place. They are not given in the pre-logicized mathematics – neither are they given in the relational framework. The delineation of their content requires that a complicated process of adjustment between mathematics and logic has been led to completion. Now, every step in this process could be challenged. One cannot thus justify the decision to draw the distinction between logical content and psychological accompaniment by referring to the identity of relational types: the way they are defined is a consequence (not a cause) of the way we draw this limit. In PoM, no general mechanism allowing us to correlate a given mathematical practice to a relational type is available. It is then at this level that the problem of demarcation between the psychological and the logical resurfaces.

3.6 Conclusion

In this chapter, I have first argued against the if-thenist interpretation. It is true that, for Russell, any axiomatic theory (which has an arithmetical model) can be regarded as a characterization of a logical structure. But this does not mean that any axiomatized theory (with a model) should be automatically regarded as a part of logical knowledge. Metrical geometry is an axiomatized science, which has an arithmetical model – and yet, metrical geometry was ultimately seen as an empirical science in PoM. Axiomatization and existence-theorem are necessary conditions for a science to be considered as a logical science. They are not however sufficient. Some further constraints need to be fulfilled as well. The question is then of course the following one: what is the nature of the supplementary constraints which confer to an axiomatized theory the status of a logical science? This issue is intimately connected with the one related to the nature of the non-formal considerations Russell used to distinguish between his different formally adequate analyses of the same concept. Here and there, the problem is the same: to choose among different formally equivalent alternatives, Russell referred to some features whose logical status was not clear.

This difficulty is brought to the fore when Russell's fierce battle against psychologism is brought into the picture. It is well-known that Russell emphasized that one should always distinguish the content of a mathematical concept or theory from our knowledge of it. In particular, this distinction was often used to dismiss certain objections to logicism. Now, it seems that nothing prevents us from turning Russell's argument against him – the considerations Russell used in PoM VI to disqualify certain formally correct analyses could be considered as the product of a mix-up between logic and psychology. How then can we draw the boundary lines between the two domains? Is there a general mechanical way to solve this problem?

My answer here has been negative. I first showed that the (neglected) notion of a relational type can help to alleviate the tension raised by the psychologist threat. The idea that mathematical topics could be defined as classes of relations provides us with a means to give a logical status to the non-formal features Russell relied on in his analyses. But this first answer seemed to side step, rather than really solve, the difficulty. The identities of the fundamental relational types are not given before the analysis begins – they are, on the contrary, reached through the analytical process. And one cannot justify a decision taken during the analytical process by referring to the final stage of the argument. It seems then that there is no general answer to Boutroux's kinds of argument. Russell, no doubt, drew a distinction between the logical and the psychological components of a mathematical theory. But he did not give any recipe for tracing the boundaries in particular cases. As a consequence, his constructions remained open to discussion: it might be the case that the theory of differential equations must be associated to a relational type; it also might be the case that projective geometry has a too important place in PoM. Russell had no definitive knock-out argument which could settle once and for all this kind of issue.

I will come back to this important and difficult question in Chapter 7. I will hold there that this apparently sad conclusion (Russell has no definitive argument to meet the threat of psychologism) is in fact good news. But before that, I will turn in the next three chapters to a completely different subject: Russell's analysis of quantity. I will follow there the same method that I have followed in the first three chapters – Chapters 4 and 5 are devoted to a detailed examination of Russell's conception in (respectively) PoM III and PM VI, while Chapter 6 is an attempt to extract some general conclusions from my comprehensive survey. After this detour, we will be in a better position to tackle the questions raised here.

4
Quantity in *The Principles of Mathematics*

Part III of PoM has not been widely studied. Certain aspects of Russell's doctrine of quantity have been touched on when discussing Russell's use of the abstraction principle[1] or when examining the genesis of PoM.[2] But, except Michell (1997; 1999), nobody, to my knowledge, has tried to present in detail Russell's doctrine. There are some reasons for that. First, even if it is quite short (40 pages), PoM III is terribly complicated. Russell introduced there several idiosyncratic notions (for instance: kind of magnitude, divisibility, stretch, relational magnitude) that are nowhere precisely defined and which almost never occur again in the book. Second, Russell himself, at the beginning of Part III, warned his reader that 'the whole of this part ... is a concession to tradition; for quantity ... is not properly a notion belonging to pure mathematics at all' (p. 158). He is here alluding to the fact that, thanks to the works of Dedekind and Cantor, there was no need, in 1903, to refer to quantity for giving an account of the theory of real numbers and real analysis.[3] Owing to these recent mathematical developments, Russell himself seemed to consider PoM III as a dispensable outgrowth, withdrawn into itself and unconnected with the rest of the book. No surprise then if scholars did not rush into this mire on the side.

The first reason I have to come back in detail to Russell's theory is thus the desire to fill a gap in the literature. Even if PoM III is not essential to the understanding of Russell's theory of real numbers, the lack of any study of this subject is annoying. Quantity was at the end of the nineteenth century a very fashionable topic – the neo-Kantian, but also Poincaré and Duhem in France and the followers of Brentano in Austria, dealt with the subject. To pass Russell's doctrine over in silence prevents us from relating him to his contemporaries.[4] But the main reason I have to look at this doctrine is that the last part of PM resumed and extended it. My main goal in this chapter is thus to pave the way for the presentation of Russell's mature theory of quantity which I will present in the next chapter. As I will explain, Russell launched in PM VI an attack against Dedekind's and Cantor's arithmetization programme of real analysis – a programme he still espoused in PoM.[5] But despite this difference, the PM doctrine of magnitude cannot be understood without resorting to PoM's early theory.

In the two first sections, I will draw a map of the area covered in PoM III, though I will not attempt to justify any of the views I will present. My aim is just to outline Russell's main theses as clearly as I can. In the subsequent sections, I will come back to some of the most challenging views defended by Russell. In section 4.3, I will focus on the distinction between quantity and magnitude, and on the correlative issue surrounding the notion of a 'kind of magnitude'. This section is not essential for the rest of my demonstration, but I have included it because it gives an original answer to a standard question in Russell's scholarship: why did Russell, who espoused a relativist view of numbers, endorse an absolutist conception of magnitude? Section 4.4 deals with his theory of measurable relational magnitude. This doctrine is fundamental, since it will be resumed in PM, and I will devote some time to explain it in detail.

4.1 Magnitude and quantity

At the end of the nineteenth century, many mathematicians[6] attempted to give an axiomatic version of Euclid's theory of magnitude. They often disagreed about how the concept of magnitude should be defined, and these disputes played an important role in the intellectual life of the period (as for instance the dispute concerning the legitimacy of non-Archimedean systems).[7] But there was a consensus on the idea that a magnitude should be regarded a combination of an additive structure and an ordinal structure.[8] In other words, in all these works, magnitude was always characterized as an ordered (a magnitude can be said to be greater or less than another of the same kind) semi-group (two magnitudes of the same kind can be added to yield another magnitude of the same kind).[9] As an example, let me take the axiomatic system Burali-Forti (1898) presented – which Russell read and studied.[10] $\langle G, + \rangle$ (where G is a set and $+$ a binary law) is a homogeneous magnitude if and only if the following eight conditions are satisfied:[11]

1.	$\forall a,b \in G,\ a+b = b+a$	Commutativity
2.	$\forall a,b,c \in G,\ a+(b+c) =$ $(a+b)+c = a+b+c$	Associativity
3.	$\forall a,b,c \in G,\ (a+c = b+c) \Rightarrow a = b$	All the elements of G are regular
4'.	$\forall a \in G,\ \exists x \in G,\ a+x = a$	Existence of the zero[12]
4''.	$\forall a \in G,\ \exists x \in G,\ a+x \neq a$	Existence of a magnitude not equal to 0
5.	$\forall b \in G,\ \forall x \in G \backslash \{0\},\ (a+b) \in G \backslash \{0\}$	

Definition of order : $\forall a,b \in G$, $a>b$ ssi $\exists x \in G\backslash\{0\}$, $a = b+x$

6. $\forall a,b \in G$, $a=b \vee a<b \vee a>b$ $\langle G, > \rangle$ is a totally ordered set

7. $\forall a \in G$, $\exists x \in G\backslash\{0\}$, $x<a$

8. If a non-empty set of homogeneous magnitude has an upper bound, then it has a least upper bound.[13]

In Burali-Forti's system, order is introduced by a definition and is not a primitive relation. $\langle G, + \rangle$ is nevertheless the positive cone of an ordered semi-group. We thus find in Burali-Forti's homogeneous magnitude the usual combination of an additive and an ordinal structure.

At the beginning of the twentieth century, the additive magnitudes of the kind I have just presented were usually called 'extensive' magnitudes, distinguished from 'intensive' magnitudes. Intensive magnitudes could not be added, but they could be arranged with respect to the greater and the lesser – they could be ordered. Intensities of pleasure and pain were the classical examples of intensive magnitude. The question concerning the possibility of measuring intensive magnitude was heatedly discussed during the second half of the nineteenth century. The then standard approach was that extensive magnitudes were the sole measurable magnitudes (Kant, for instance, defended such a view).[14] But the progress of physics (notably the emergence of thermodynamics[15]) and the birth of psychophysics[16] challenged this divide: some magnitudes, traditionally regarded as intensive, began to be measured. Yet, by the end of the century, extensive magnitudes were still considered as the paradigmatic case for measurable magnitudes. This fact, that all the various axiomatic systems which then blossomed focused on the additive case, is an illustration of the prominence of extensive magnitude.[17]

In PoM, Russell espoused the so-called ordinal view, according to which additivity is not essential to quantity. In this perspective, magnitude should first and foremost be defined as an ordinal structure. Russell thus wrote, at the beginning of the first chapter of PoM III (p. 159):

In fixing the meaning of such a term as *quantity* or *magnitude*, one is faced with the difficulty that, however one may define the word, one must appear to depart from usage. This difficulty arises wherever two characteristics have been commonly supposed inseparable which, upon closer examination, are discovered to be capable of existing apart. In the case of magnitude, the usual meaning appears to imply (1) a capacity for the relations of *greater* and *less*, (2) divisibility. Of these characteristics, the first is supposed to imply the second. But as I propose to deny the implication, I must either admit that some things which are indivisible are magnitudes, or that some things which are greater or less than others are not magnitudes. As one of these departures from usage is unavoidable, I shall choose the former, which I believe to be the less serious. A magnitude, then, is to be defined as anything which is greater or less than something else.

Of course, extensive magnitudes remained magnitudes; but they were just a species of a more general genus, and they should not then be taken as the paradigm of magnitude.[18] Michell (1997; 1999) has laid great stress on this point. He portrayed Russell as the first scientist to have challenged the prominence of the extensive paradigm. It should be remarked however that Russell, in PoM, maintained that only extensive magnitude could be measured (see section 4.2 below). His definition was then not the radical break described by Michell. Russell did not hold that intensive magnitudes were measurable – he just emphasized that some quantities were not measurable. But this had already been acknowledged by Kant.

Having redefined magnitudes in terms of order, Russell then introduced, in PoM chapter 19, a crucial distinction between magnitude and quantity (ibid.):

> It might be thought that *equality* should be mentioned, along with greater and less, in the definition of magnitude. We shall see reason to think, however – paradoxical as such a view may appear – that what can be greater or less than some term, can never be equal to any term whatever, and vice versa. This will require a distinction, whose necessity will become more and more evident as we proceed, between the kind of terms that can be equal, and the kind that can be greater or less. The former I shall call *quantities*, the latter *magnitudes*. An actual foot-rule is a quantity: its length is a magnitude. Magnitudes are more abstract than quantities: when two quantities are equal, they have the *same* magnitude. The necessity of this abstraction is the first point to be established.

Two elements a and b of a totally ordered set (of a quantitative structure) can always be compared so as to generate one of the three following situations: (1) a is greater than b; (2) a is less than b; (3) a is equal to b. If the quantitative order is a non-strict ordering, then the third situation corresponds to the case where one has both $a \geq b$ and $b \geq a$. This relation is an equivalence relation, and the set of quantities is then partitioned by the relation of equality in magnitude. Russell, however, did not maintain this reasoning. He distinguished two correlated ordinal relations which apply to two correlated sets of entities: the quantities and the magnitude. To the quantities a and b he associated magnitudes A and B in such a way that if a is greater (smaller) than b, then $A > B$ ($A < B$) and that if a is quantitatively equal and b, then $A = B$. In other words, if A and B are two different magnitudes of a given kind, one can have either $A > B$ or $B > A$ – but, unlike what happened with quantities, two different magnitudes can never be equal: 'when two quantities are equal, they have the *same* magnitude'.[19]

This distinction between two correlated series of entities, quantities and magnitudes, is surprising. Russell knew very well that a non-strict total ordering always induces a strict total ordering of the equivalence classes of equal quantities. Why did he not hold, then, that magnitudes are nothing else than the equivalence classes of equal quantities? This is especially puzzling because, as we saw in Chapter 3, Russell identified cardinal numbers with the equivalence classes of similar sets. Why maintain an 'absolutist' view of quantity, while espousing a 'relativist' conception of number? I will come back to this in section 4.3. For the moment, let us content ourselves with noting that Russell differentiated two distinct correlated series: quantities and magnitudes.

In Russell, the orderings are always regarded as total – that means that whichever a and b are chosen (whichever A and B are chosen), we will have $a \geq b$ or $b \geq a$ ($A > B$ or $B > A$ or $A = B$). This compels us to introduce some restrictions on the set of magnitudes. Imagine that a is a segment and b is a certain rectangular area in the Euclidean plane. The two entities a and b are quantities: I can compare the segments and I can order rectangular areas. But I cannot, in any natural way, compare a segment (a length) to a rectangle (an area); a segment is neither less nor greater than a rectangle; the two are incommensurable. Now, since the quantitative order is always total, a and b (A and B) cannot belong to the same set of quantities (of magnitudes). To block this kind of difficulty, Russell introduced in 1903 the notion of a kind of quantity (magnitude). In PoM, quantities and magnitudes are divided up into different and incompatible kinds. Within any of these kinds, any two elements can be quantitatively compared; but no comparison across kinds is possible.

The idea that there are different[20] kinds of magnitude, among which no cross-ordering is allowed, is not as simple as it appears at first. This division of the quantitative world seems obvious when certain examples are considered (as length and area, for instance); but other situations raise more serious difficulties. Pleasure is an especially difficult case. Is it really a kind of magnitude? Or does one have to distinguish several distinct kinds (different qualities) of pleasure? This issue, which played a key role in the criticism Mill addressed to Bentham's version of utilitarianism, is discussed at length by Moore (1903, §§ 46–8). Russell also alluded to the question in PoM.[21] More generally, how can we justify this prohibition of cross-kind comparisons? It seems that there are two possibilities. One could first hold that the relation of quantitative comparison determines the boundaries of the different kinds – as a matter of fact, one cannot compare a segment and a rectangle, and this fact would explain why one should distinguish the kind of length from the kind of area. But one could also consider that the notion of kind comes first and determines the possibilities of comparisons – it would be because segments and areas belong to different kinds that one could not compare them. Which is first: the kind or the operation of quantitative comparison? As we will see in section 4.3, Russell espoused in PoM III the realist view: there are kinds of magnitude, and these kinds ground the possibilities of quantitative comparisons. A quantity (a magnitude) was always, for Russell, a quantity *of* something (its kind: length, area, pleasure, etc.). But for the moment, let us note the new idea he introduced: in addition to the distinction between quantity and magnitude, there is a difference between kinds.

To summarize, in PoM chapter 19, Russell presents a three-levelled building: quantities fill the ground level; at level one, there are magnitudes, which are the common properties of equal quantities; at level two, there are kinds of magnitude, which are the properties possessed by comparable magnitudes (and by way of consequences, possessed by comparable quantities). Chapter 20 complicated this schema in two ways. First of all, Russell, continuing some ideas contained in Meinong (1896) (for more on this, see pp. 119–20), claims that certain magnitudes are relational. As an example, he took the degrees of resemblance between shades of red. Let us assume that b, c and a are three shades of red and that b resembles a more than c; in this case, the resemblance between b and a (bRa, let's say) is greater than the resemblance between c and a (cSa) – the relational magnitude R instantiated by b and a can in this way be quantitatively compared to the relational magnitude S instantiated by c and a. This example shows that magnitudes are not always properties of quantities; relations can also have a magnitude.[22] Russell applies to the relational case the distinction between quantity and magnitude. Let's suppose that the degree of resemblance between two shades of red a and a' is the same as the degree of resemblance between two other shades of red b and b'. If Q is the relation of resemblance between a and a', one will then have aQa' and bQb' – or in other words Q will be the common magnitude of two quantities $a'Qa$ and $b'Qb$. In the relational case, magnitudes are relations R, S, Q, etc. belonging to a common kind, while quantities are the various instances of these relations bRa, cSa, $a'Qa$, $b'Qb$, etc. The relational quantities having the same magnitude are thus not differentiated by their spatio-temporal positions, but by the terms between which the relational magnitude holds. As Russell wrote (PoM, p. 167): 'when a magnitude can be particularized by temporal, spatial, or spatio-temporal position, or when, *being a relation, it can be particularized by taking into consideration a pair of terms between which it holds*, then the magnitude so particularized is called a *quantity*' (my italics). In the case of relational magnitude, one again finds the three levelled structure:

Level 2: **Type** of relational magnitudes.

Level 1: Relational **magnitudes** which hold or do not hold between certain terms.

Level 0: Terms to which the relations of level 1 apply – and which differentiate the various **quantities** associated with a same relational magnitude.

There is a second feature which complicates Russell's account. It has to do with the issue of divisibility we have already mentioned. In chapter 20, Russell holds that a magnitude is never divisible (PoM, p. 173):

All the magnitudes dealt with hitherto have been, strictly speaking, indivisible. Thus the question arises: Are there any divisible magnitudes? Here I think a distinction must be made. A magnitude is essentially one, not many. Thus no magnitude is correctly expressed as a number of terms.

This seems, at first, strange. Indeed, we have already seen that, at the beginning of the twentieth century, measurement was intimately connected to additivity, and then to divisibility. If magnitudes are never divisible, how can we account for the fact that we can measure length, mass, time, etc.? To understand Russell's idea, one has to bring in the distinction between magnitude and quantity. The passage just quoted goes on as follows (ibid.):

Thus the question arises: Are there any divisible magnitudes? Here I think a distinction must be made. A magnitude is essentially one, not many ... But may not the quantity which has a magnitude be a sum of parts, and the magnitude a magnitude of divisibility? ... But though the whole which has divisibility is of course divisible, yet its divisibility, which alone is strictly a magnitude, is not properly speaking divisible. The divisibility does not itself consist of parts, but only of the property of having parts. It is necessary, in order to obtain divisibility, to take the whole strictly as *one*, and to regard divisibility as its adjective. Thus although, in this case, we have numerical measurement, and all the mathematical consequences of division, yet, philosophically speaking, our magnitude is still indivisible.

Holding that a magnitude is never divisible does not amount to saying that the 'concrete' quantities instantiating it are not divisible. When the magnitude in question is a property,[23] the spatio-temporal quantities correlated to these magnitudes can be divided up and added to each other. In the extract quoted above, Russell explains that, in certain cases, there is a relation between the magnitudes and the degrees of 'divisibility' of the quantities. Take two divisible quantities a_1 and a_2 of equal magnitude A; it could happen that the quantity b, which results from the 'addition' of a_1 and a_2, has a magnitude B twice as great as A. When this occurs, the magnitude in question is said to be a magnitude of divisibility. The magnitudes A and B are not themselves divisible; they cannot be joined together, for instance to form a new magnitude. But an additive structure between magnitudes can nevertheless be derived from the properties of the addition between quantities. I will come back to this idea very soon. What is important for now is to understand that the thesis that magnitude is indivisible does not imply that there are no extensive (and so measurable) magnitudes. It only means that the additive operation between magnitudes cannot be introduced as a primitive operation. In other words, the claim that magnitude is indivisible does not preclude the construction of a theory of measurement – it only makes it more difficult.

4.2   The concept of a measurable magnitude

Russell tackles the issue of measurement in chapter 21, entitled 'Numbers as Expressing Magnitudes: Measurement'. Let me quote Russell's introduction (PoM, p. 176):

> Measurement of magnitudes is, in its most general sense, any method by which a unique and reciprocal correspondence is established between all or some of the magnitudes of a kind and all or some of the numbers, integral, rational, or real, as the case may be ... In this general sense, measurement demands some one–one relation between the numbers and magnitudes in question – a relation which may be direct or indirect, important or trivial, according to circumstances. Measurement in this sense can be applied to very many classes of magnitudes; to two classes, distances and divisibilities, it applies, as we shall see, in a more important and intimate sense.
>
> Concerning measurement in the most general sense, there is very little to be said. Since the numbers form a series, and since every kind of magnitude also forms a series, it will be desirable that the order of the magnitudes measured should correspond to that of the numbers, i.e. that all relations of *between* should be the same for magnitudes and their measures. Wherever there is a zero, it is well that this should be measured by the number zero. These and other conditions, which a measure should fulfill if possible, may be laid down; but they are of practical rather than theoretical importance.

According to Michell (1999, p. 118), Russell would here define measurement as 'a one-to-one correspondence between the magnitudes of a quantity and a subset of one or other of the number systems, integral, rational or real, in such a way that the order of the magnitudes is represented by the order of the corresponding numbers' – and this would constitute 'the first explicit statement of the representational view of measurement'. I agree with Michell that Russell gave in the passage quoted above a rough sketch of a 'representational' approach to measurement. But this general characterization did not play any role in PoM. As Russell noted, about 'this most general sense, there is very little to be said'. And in fact, the rest of chapter 21 is entirely devoted to a more particular sense of measurement, the only significant one for Russell, which applied 'in a more important and intimate sense' to two kinds of magnitude, 'distance and divisibilities'. My aim in this section is to explain this more particular and more important doctrine (neglected in Michell 1999).

Let me first recall the problem. A class of extensive magnitude is an ordered set endowed with a binary operation of addition. Now, Russell holds that no magnitude is by itself divisible. This seems to amount to denying that addition could be introduced as an operation. How then can we account for the fact that certain magnitudes are extensive? Chapter 21 aims at solving this puzzle, that is, at explaining how we can give 'an intrinsic meaning to the proposition: "this magnitude is double of that"' (PoM, p. 178), without giving up the idea that magnitudes are indivisibles. To be more specific, Russell presented two distinct solutions to this problem: the theory of magnitude of divisibility and the theory of relational distance. I will examine each of these in turn.

4.2.1   The theory of magnitude of divisibility

The theory of magnitude of divisibility (which we have already encountered in Chapter 2, when dealing with the empirical view of metrical distance) is a difficult theory that is never fully worked out in PoM. It is also something that Russell abandoned soon after 1903 (see note 28) and that did not play any role in PM. Owing to this, I will content myself here with giving a brief sketch of Russell's theory and of its difficulties. Russell explains the core of the theory in the following way (PoM, p. 178):

> So long as quantities are regarded as inherently divisible, there is a per-fectly obvious meaning to [the proposition: 'this magnitude is double of that']: a magnitude A is double of B when it is the magnitude of two quantities together, each of these having the magnitude B.

Consider a segment a of the Euclidean line. This is a quantity that Russell, in PoM, called a stretch. The stretch a has a certain length, the magnitude A. Now, as a Euclidean segment is a divisible quantitative whole, one can divide the stretch a into two equal stretches b_1 and b_2. Designating the concrete operation of lying end to end the two quantities b_1 and b_2 by '#', one can then write that $a = b_1 \# b_2$. Russell suggests that, in this case, the magnitude A of a is the double of the magnitude B, common to b_1 and b_2 – that is, he holds that $A = B + B$. The main idea is then that, in certain cases, the properties of the operation of 'concatenation' between quantities allow us to induce an operation of addition between the magnitudes correlated, even though these magnitudes cannot really be added. When this happens, we say that A, B, C, etc. are magnitudes of divisibility: the more divisible the quantities are, the greater the correlated magnitudes become.

The doctrine was in many respects unclear. The main problem concerned the concept of divisibility. As Russell explained in § 167 (and also in § 147 and § 397), there are two distinct sorts of quantitative wholes: the wholes which have a finite number of terms, and those which have an infinite number of terms. Let a and b be two distinct finite quantitative wholes; then $a\#b$ can be thought of as a new whole which is the set-theoretical union of a and b. Now, the magnitudes of a, b and $a\#b$ (annotated respectively as $\text{Mg}(a)$, $\text{Mg}(b)$, $\text{Mg}(a\#b)$) can in this case be identified with the cardinal of these classes – and as a and b have no element in common,[24] we will have:

$$\text{Mg}(a\#b) = \text{Card}(a \cup b) = \text{Card}(a) + \text{Card}(b) = \text{Mg}(a) + \text{Mg}(b).$$

But when a and b are two distinct infinite wholes (for instance, two segments of the Euclidean line), then the same reasoning would give:

$$\text{Mg}(a\#b) = \text{Card}(a \cup b) = \text{Card}(a) = \text{Card}(b) = 2^{\aleph_0} = \text{Mg}(a) + \text{Mg}(b) = \text{Mg}(a) = \text{Mg}(b).[25]$$

This cannot be, since the divisibility of $a\#b$ should be strictly greater than the divisibility of a. In the infinite case (which is the most interesting one), the concatenation of quantitative wholes cannot be seen as a set-theoretical union, and divisibility cannot be defined as cardinality. How, then, can we characterize it?

Russell never answered the question. At some places (PoM, pp. 411–12), he suggested defining the magnitude of divisibility as an 'ordinal notion' – but he quickly acknowledged that this would not answer the difficulty either.[26] No other attempt is made to explain the nature of the divisibility of an infinite quantitative whole. What we find instead is a development about our capacity to estimate whether two infinite wholes are equal. Thus (PoM, p. 178):

> In the case of infinite wholes ... the number of simple parts ... may be equal without equality in the magnitude of divisibility. We require here a method which does not go back to simple parts. In actual space, we have immediate judgments of equality as regards two infinite wholes. When we have such judgments, we can regard the sum of n equal wholes as n times each of them; for addition of wholes does not demand their finitude. In this way numerical comparison of some pairs of wholes becomes possible. By the usual well-known methods, by continual subdivision and the methods of limits, this is extended to all pairs of wholes which are such that immediate comparisons are possible. Without these immediate comparisons, which are necessary both logically and psychologically, nothing can be accomplished.

We saw in Chapter 2 that Russell used this idea to counter Poincaré's conventionalist arguments. But, from our actual perspective, Russell's argument is nothing but an escape. To claim that humans can make judgements about divisibility is one thing – to define the notion of divisibility is another one. And Russell never provided an adequate characterization of the divisibility of an infinite whole. This is puzzling because, to fulfil this task, he could have relied on the concept of a measure (or 'content') of a set that many mathematicians at the time were investigating. He could have, for instance, made use of the notion of a 'measure function' which surfaced in some of Jordan's and Peano's works.[27] But, surprisingly, the (at the time) growing theory of measure is completely absent from PoM, and Russell did not draw any connection between his conception of divisibility and the idea of measure.[28] As a consequence, his views of divisibility never reached the conceptual strength of his development about order or about cardinality. No surprise, then, if he gave up the theory quickly after 1903.

4.2.2 The theory of relational distance

Let me now turn to the second and more interesting way of giving 'an intrinsic meaning' to addition between indivisible magnitudes. Russell developed his analysis in § 168 of chapter 21 of PoM (see also pp. 252–5). Here is how he introduces the subject (pp. 179–80):

> [In the case of magnitudes of divisibility] we still had addition on one of its two fundamental senses, i.e. the combination of wholes to form a new whole. But in other cases of magnitude we do not have any such addition. The sum of two pleasures is not a new pleasure, but is merely two pleasures. The sum of two distances is also not properly one distance. But in this case we have an extension of the idea of addition. Some such extension must always be possible where measurement is to be effected in the more natural and restricted sense which we are now discussing.

There are some indivisible quantities, said Russell, which can be measured. The doctrine of divisibility should then be complemented by another one, applicable to these sorts of magnitudes. In the passage, he refers to the example of distance. At first sight, this seems awkward: since Euclid, a segment is the paradigmatic case of divisible magnitude. But Russell's notion of a distance is very peculiar. First, 'the word [distance] will be used here to cover a far more general conception than that of distance in space' (PoM, p. 180); for instance, two shades of red are related by a distance. Second and more important, distance is here a relation, not a spatio-temporal whole. Thus, the distance between two points *a* and *b* is not the segment [*a*, *b*] – it is the particular relation that the two extremities of the segments have. The distance between *a* and *b* is not a set of points; it is a relation, the geometrical transformation sending *a* to *b*. As one cannot divide relation, a distance is not a magnitude of divisibility. How then can one measure it? How could one introduce an addition between relational distances? In the rest of § 168, Russell explains his idea (p. 180):[29]

> I shall mean by a kind of distance a set of quantitative asymmetrical relations of which one and only one holds between any pair of terms

of a given class; which are such that, if there is a relation of the kind between *a* and *b*, and also between *b* and *c*, then there is one of the kind between *a* and *c*, the relation between *a* and *c* being the relative product of those between *a* and *b*, *b* and *c*; this product is to be commutative, i.e. independent of the order of its factors; and finally, if the distance *ab* be greater than the distance *ac*, then, *d* being any other member of the class, *db* is greater than *dc*. Although distances are thus relations, and therefore indivisible and incapable of addition proper, there is a simple and natural convention by which such distances become numerically measurable.

This passage is extremely difficult. Apart from the fact that distances are relations, it is hard to extract from it any intelligible piece of information. However, unlike what happened in the case of the magnitude of divisibility, Russell's ideas are both extremely clear and very important to grasp in relation to what happened in PM. As we will see, what makes the extract difficult to understand is the use of ordinary English. Fortunately, Russell gave a formal presentation of the theory (1900b), published for the first time in volume III of the *Collected Papers*. The key idea of his construction, more apparent in (1900b), is to define addition as a composition (relative product) of relations. When some conditions on this composition operation are satisfied (these conditions are the ones listed in the passage of PoM above), then the addition has all the properties required to make the magnitudes measurable. The additive structure is thus not generated by a concatenation operation between spatio-temporal quantities, but introduced via the logical operation of the composition of relations. As the subject is crucial for what comes next, I will devote section 4.4 to the study of Russell (1900b).

Let me summarize what we have seen so far by quoting the conclusion of chapter 21 (PoM, p. 182):

We thus see how two great classes of magnitudes – divisibilities and distances – are rendered amenable to measure. These two classes practically cover what are usually called extensive magnitudes, and it will be convenient to continue to allow the name to them. I shall extend this name to cover all distances and divisibilities, whether they have any relation to space and time or not. But the word *extensive* must not be supposed to indicate, as it usually does, that the magnitudes so designated are divisible. We have already seen that no magnitude is divisible. *Quantities* are only divisible into other quantities in the one case of wholes which are quantities of divisibility. Quantities which are distances, though I shall call them extensive, are not divisible into smaller distances; but they allow the important kind of addition explained above, which I shall call in the future *relational* addition [or relative product].

We have seen that Michell (1997; 1999) defended the idea that Russell's theory of measurement anticipated the representational approach of Suppes and his followers. But it seems that Michell misplaced the originality of Russell's approach. Russell did not aim at challenging the centrality of extensive magnitude; on the contrary, he still identified extensive with measurable magnitude. What was original, however, in PoM III, is the way he broke up the genus 'extensive magnitude' into two different species: the magnitudes of divisibility and the relational distances. He actually borrowed the idea from a neglected book from Meinong (1896). In the review of this work he did in 1899, Russell praised Meinong for being the first to set out a contrast between two kinds of geometrical extensive magnitude, the relational distances (*Distanz*) and the divisible lengths (*Strecke*). Thus (1899d, p. 252):

> The first section of [Meinong (1896)] consists of a discussion of the nature and range of quantity. It is pointed out that quantities need not be divisible, since relations may be quantities. Distance in space, for example, is unquestionably a quantity and a relation: to suppose distance is divisible, can only arise from a confusion between distance and length (*Strecke*). In like manner, the author continues, similarity and dissimilarity are quantities: two things may be more or less similar, but the similarity is certainly indivisible.

As Russell explained, this difference between distance and length is for Meinong a particular application of a general distinction between two kinds of extensive magnitude: the relational *Verschiedenheit* (dissimilarity) and the divisible *Differenz* (difference).[30] One finds here the origin of the idea that extensive magnitude should be broken up into two subspecies, distance and magnitude of divisibility. Surprisingly enough, Russell did not allude in PoM to the Meinongian pedigree of his doctrine. The philosopher seemed more interested in underlining the contrasts between himself and Meinong[31] than acknowledging his debts (the influence of this idea is yet explicitly recognized in some places, e.g. PoM (pp. 252–3)).[32]

Anyway, this division of extensive magnitude into two distinct subspecies is a very important move, since it played a crucial role in Russell's philosophy of metrical geometry. As we have seen in Chapter 2, two (in fact, three, but the third one was downplayed) different conceptions of metric were developed in PoM: the projective notion, in which metric is construed as a relational magnitude, and the empirical view, in which geometrical distance is seen as a magnitude of divisibility. This approach, which was deeply rooted in the mathematics of the time (especially in Klein's work, see pp. 73–4), was also anchored in Russell's 'Meinongian' theory of extensive magnitude. The idea that there really are two distinct notions of linear distance, always confused by philosophers, follows from Russell's analysis of measurable quantity.[33] There is thus a strong consistency between PoM III and PoM VI: the theory of metrical distance resumes the fundamental distinctions expounded in the general theory of measurable magnitude; conversely, the abstract analysis of magnitude finds support in Russell's detailed analysis of metrical geometry.

4.3   The absolutist theory and the concept of a kind of magnitude

Let me thus summarize the intricate structure of Russell's 1903 doctrine of magnitude. In chapter 19, he defined quantity and magnitudes as an ordered set. He then made a distinction between magnitude and quantity, and explained that all quantities (all magnitudes) of the same kind have a relation of greater and less – but that no quantitative comparison across kinds is possible. In chapter 20, he added that there are relational magnitudes, and that a magnitude (unlike a quantity) is always indivisible. This last claim raised a problem: how can we account for extensive magnitude? Chapter 21 gave a twofold answer to this question. Addition between magnitudes can firstly be derived from the operation of concatenation between the quantitative wholes associated to the magnitudes. It can secondly be derived from the relative product between relational magnitudes. This last doctrine is important since it is the matrix of Russell and Whitehead's mature theory of magnitude presented in PM VI. Before explaining Russell's theory in more detail, I would like to come back to the distinction between quantity and magnitude.

Russell held in PoM III an 'absolutist' conception of magnitude, according to which magnitudes were not reducible to equivalence classes of equal quantities. The same Russell supported in PoM II a 'relativist' view of the cardinals, according to which numbers should be identified with equivalence classes of similar classes. How can we account for this discrepancy?[34] Scholars usually relied on the complicated genesis of PoM to explain Russell's versatility on this matter. According to this interpretation, the absolutist theory of magnitude would pertain to an old stratum of thought that Russell did not recover by the time of more recent developments before publishing the book. I will claim here, against this view, that he had some deep reasons to develop an absolutist conception of magnitude while espousing a relativist view of number.

Let me recall the key features of Russell's analysis of cardinal numbers. As I explained in Chapter 3, Russell, in PoM, based his reasoning on a theorem (the theorem of abstraction), first published in Russell (1901a), which states that R being an equivalence relation, 'there is some [many–one] relation S, such that, if xRy, there is some one entity t for which xSt, ySt' (PoM, p. 221). Before this date, he referred to a principle (the principle of abstraction), which stipulated that, when an equivalence relation R is given, one must admit that the terms having this relation have a common property or content. The import of Russell (1901a) is to show that this common property is not necessarily unique. This is a feature on which he strongly insisted in Russell (1901a),[35] and which he used in PoM to criticize Peano's definition (p. 114; see also pp. 81–2 and 87–8 in this book). Cardinal numbers cannot be characterized as the properties common to all similar sets, since there are many such properties. To disambiguate Peano's definition, Russell chose to 'define a number of a class as the class of all classes similar to a given class' (PoM, p. 115). This amounted to identifying S as the relation of class-membership and t as an equivalence class, and then holding a relativist conception of cardinality.

Fifty pages after having presented his relativist analysis of cardinals, Russell refused to identify a magnitude with a collection of equal quantities. How can we explain his decision? The mystery deepens, since, in PoM III, he referred to the 'principle of abstraction' to justify his absolutist conception of magnitude (p. 166). Furthermore, in a footnote added in 1902,[36] he wrote that 'a magnitude may, so far as formal arguments are concerned, be identified with a class of equal quantities' (PoM, p. 167). Russell recognized then the analogy between the two cases. Why did he then not do for magnitude what he did for cardinal? A natural reaction is to invoke the difficult genesis of PoM. Thanks to the publication of Russell's manuscripts and the works of Byrd, one now knows that PoM is in reality a patchwork of texts written at different periods. In particular, PoM III was finished in 1900, while PoM II was reworked by Russell up until 1902. So the doctrine of magnitude preceded by more than two years the theory of cardinal numbers, and during the interval he developed his new relational theory of abstraction (Russell 1901a). From this perspective, the note added in 1902 (the one where he recognized that the formal strategy used in Part II could have been carried out in Part III) would represent an awkward attempt, made at the very last minute, to relate the two analyses.

This answer is forcefully developed by Levine (1998; 2009). Levine holds that Russell, after a first idealist period (R_1), successively espoused two different metaphysical conceptions. After the *Congrès de Paris* and the acceptation of the Cantorian infinite in 1900, Russell adopted the logicist reductionism (R_3). But before that, from 1898 to 1900 (roughly), he defended a version of Moorean Platonism, according to which there was an intelligible world, made of some independent series – numbers, time, spatial positions, magnitudes, etc. – whose elements have being, but no existence (R_2). Levine argues that certain elements in R_3 (the definition of numbers and Russell's criticism of Peano's definition by abstraction, especially) ran against the fundamental tenets of R_2. He also holds that some parts of the published PoM were written during R_2, and that Russell, for whatever reason, never took the trouble to rewrite them.[37] The distinction between quantity and magnitude would be, in this perspective, only a remnant of R_2, a stage of thought which Russell had overcome at the time of the publication of the book. Even if I find Levine's interpretation attractive, I do not share the idea that Russell would have no conceptual reason to keep with his absolutist conception of magnitude. There is no doubt that the doctrine expounded in PoM III comes from a very early conceptual layer, but I believe that some deep philosophical considerations led him to maintain his former absolutist theory within the renewed framework. Recall that, when presenting his definition of numbers, he warned us that his approach, which could formally be taken each time we are confronted with an equivalence relation, should not be followed blindly. Thus, he wrote that 'if we can find, by inspection, that there is a certain class of predicates, of which one and only one applies to each collection of similar classes, then we may, if we see fit, call this particular class of predicates the class of numbers' (PoM, p. 116). In the case of cardinals, he explained that we do not have such a direct intuition. My suggestion is that, in the case of magnitude, the situation is different.

There is indeed a fundamental difference between numbers and magnitudes. As Frege (1884, § 24) noted, numbers apply to everything. One can count concrete objects and ideas, goats and meters, angels and Gods, satellites and thoughts. Numbers can be indifferently attached to any sortal concept and to any class. Numbers have no kind: a number is the same, whatever kinds of entities it is ascribed to. By contrast, any quantitative statement is always relativized to a certain kind of entity. A particular magnitude (like a particular quantity) is essentially, for Russell, a magnitude *of* something (pleasure, mass, length, etc.). This division of the sphere of magnitudes into non-overlapping kinds has thus no equivalent in the case of cardinal numbers. My suggestion is that taking into account this asymmetry could allow us to account for the difference in treatment between number and magnitude. Russell defended a realist view of the kinds of magnitude – could one not take this as the source of his defence of the absolutist conception of magnitude?

At first sight, the two theses seem to be completely independent: that the range of quantities is divided up in different exclusive kinds does not appear to imply that magnitudes exist independently of quantities. I do think, however, that there is a link between the two claims. Russell discussed the relation between kinds and magnitudes in two different places, first in PoM chapter 19 (pp. 164–5) and then at the end of chapter 20 (pp. 182–3). Both passages are difficult, but the main idea is clear enough: a particular magnitude of a certain kind, let's say pleasure, is not a simple notion and not a combination of a kind and a magnitude. Thus Russell wrote (pp. 164–5):

> A magnitude of pleasure is so much pleasure, such and such an intensity of pleasure. It seems difficult to regard this ... as a simple idea: there seem to be two constituents, pleasure and intensity. Intensity need not be intensity of pleasure, and intensity of pleasure is distinct from abstract pleasure. But what we require for the constitution of a certain magnitude of pleasure is, not intensity in general, but a certain specific

intensity; and a *specific* intensity cannot be indifferently of pleasure or of something else. We cannot first settle how much we will have, and then decide whether it is to be pleasure or mass. A specific intensity must be of a specific kind. Thus intensity and pleasure are not independent and coordinate elements in the definition of a given amount of pleasure ... Thus *magnitude of pleasure* is complex, because it combines magnitude and pleasure; but a particular magnitude of pleasure is not complex, for magnitude does not enter into its concept at all ... This is more easily understood where the particular magnitude has a special name. A yard, for instance, is a magnitude, because it is greater than a foot; it is a magnitude of length, because it is what is called *a* length.

In the Aristotelian definition by genus and differentia, the concept 'human' (let's say) appears as a combination of two independent elements, the class-concept 'biped' (the genus) and the class-concept 'featherless' (the differentia). At first, one could think that this model also applies to the notion of a magnitude of pleasure – that is, one could hold that a magnitude of pleasure is a conjunction of two separate properties, pleasure and magnitude. But this does not work: 'we cannot first settle how much we will have, and then decide whether it is to be pleasure or mass'. Intensity is not an independent concept, which can be conjoined or not to a kind of pleasure; a magnitude of pleasure is just a certain amount of pleasure; and conversely, pleasure comes always at a certain degree. One cannot have intensity without kind; one cannot have kind without intensity.[38]

What Russell is trying to say here is that the relation between kind and magnitude falls within the province of what the philosopher and logician W. E. Johnson (1921)[39] called the relation of a determinable and determinate. A. N. Prior (1949, p. 6) described Johnson's theory thus:

> What is coloured is thereby 'determinable' as red or blue or whatever it may be, and so we might call its colour a determinable part of its nature, pars *determinabilis essentiae*; Johnson calls it, simply, a 'determinable'. But the correlative to this term, in Johnson, is not, as in the Schoolmen, 'determinant', but 'determinate'; and this is not just a verbal slip. For redness, say, is not a pars *determinans essentiae* or differentia which by combination with the determinable 'colour' produces some new complex quality; it is rather itself the *essentia determinata*, the specific nature defined. For what could we define as 'coloured and red' except 'the red'? The point about determinable genera and determinate species, in Johnson's sense of 'determinable' and 'determinate', is that there is no distinct differentia employed in passing from the former to the latter.

A determinate is not a conjunction of a genus and a differentia, but a determination of the genus. As Johnson said, a determinable 'generates'[40] its own determinates – the determinates are included in their determinable, and one cannot grasp the determinable independently of its determinations. In other words, the relation between a kind and its magnitudes, like the relation between a determinable and its determinates, is an internal relation. One cannot separate the identity of the terms from the relation which holds between them. A given magnitude is essentially a magnitude of something; and conversely, a kind is nothing except what different intensities have in common. That the relation between magnitude and kind is an internal relation is nowhere explicitly acknowledged in PoM (though it is explicitly vindicated by Johnson) – but Russell comes close to admitting it when he said that the relation between kind and magnitude 'is very peculiar, and appears to be incapable of further definition' (PoM, p. 164), that it is '*sui generis*' and should not be 'identified with the class-relation' (PoM, p. 187).

Now, what is the link between this development and Russell's absolutist theory of magnitude? Russell draws the connection at the end of chapter 20. As the passage is difficult, let me quote it in full (pp. 174–5):

> Let us start with Bentham's famous proposition: 'Quantity of pleasure being equal, pushpin is as good as poetry' ... If we suppose the magnitude of pleasure to be not a separate entity, a difficulty will arise. For the mere element of pleasure must be identical in the two cases, whereas we require a possible difference of magnitude. Hence we can neither hold that only the whole concrete state exists, and any part of it is an abstraction, nor that what exists is abstract pleasure, not magnitude of pleasure. Nor can we say: We abstract, from the whole states, the two elements magnitude and pleasure. For then we should not get a quantitative comparison of the pleasures. The two states would agree in being pleasures, and in being magnitudes. But this would not give us a magnitude of pleasure ... Hence we cannot abstract magnitude in general from the states, since as wholes they have no magnitude. And we have seen that we must not abstract bare pleasure, if we are to have any possibility of different magnitudes. Thus what we have to abstract is a magnitude of pleasure as a whole. This must not be analysed into magnitude and pleasure, but must be abstracted as a whole. And the magnitude of pleasure must exist as a part of the whole pleasurable states, for it is only where there is no difference save at most one of magnitude that quantitative comparison is possible.

Russell aims here at showing that the relativist is led, at best, to define a magnitude of pleasure as a conjunction of magnitude and pleasure. His key remark is that relativists must not only explain how they can abstract magnitude from quantities, but that they should account as well for the notion of a kind (the notion of pleasure). Let's suppose, with the relativist, that magnitude of pleasure 'is not a separate entity' and that there are only quantities of pleasures. By definition, to be comparable, two wholes must pertain to the same kind, though they should not necessarily be quantitatively equal. As Russell said, 'the mere element of pleasure must be identical in the two cases, whereas we require a possible difference of magnitude'. The problem is that, in order to abstract the notion of a pleasure, one must ignore the differences in degree between the concrete quantitative wholes. And conversely, if one focuses on the quantitative differences between pleasures, one is led to lose sight of the common property that they all exemplify. In the former case, one has the kind without the magnitudes; in the second one, the differences of magnitude without the kind. To correct this, one could try to abstract, 'from the whole states, the two elements magnitude and pleasure'. But one would then fall back into the difficulty noted before: 'the two [quantitative wholes] would agree in being pleasures, and in being magnitudes', but 'this would not give us a magnitude of pleasure'. A magnitude of pleasure is not a combination of magnitude and pleasure.

Russell's official doctrine is that relations are always external. In a certain sense then, both the absolutist and the relativist approaches failed to account for the special and *sui generis* internal relation there is between a magnitude and its kind.[41] But the absolutist can at least recognize the fact that a kind of magnitude is a 'very peculiar' concept, while the relativist is committed to the view that a kind of magnitude is a combination of two characters, both abstracted from the concrete quantities. Note that this difficulty does not arise with respect to cardinal numbers. As I have emphasized above, numbers have no kind. As only one abstraction is needed in this case, the question relative to the coordination of two distinct abstractive processes (the abstraction of the magnitude and the abstraction of the kind) does not come up.

To summarize my point, let me quote the insightful diagnosis Bigelow and Pargetter (1988, p. 288) make:

In explaining the nature of quantities, we are drawn to the core of traditional metaphysics. The theory of universals (that is, of properties and relations) arises from a recognition that, in some sense, two distinct things may be both 'the same', and 'different', at the same time ... The conflict evaporates, however, when we say that the things which are the same *in one respect* may yet be different *in other respects*. Different properties or relations constitute those different 'respects'. Quantities cause problems, because it seems as though two things may be both 'the same' and 'different' – *in the very same respect*. Two things may be the same, in that both *have* mass; yet they may be different, in that one has *more* mass than the other.

The relativist approach is at ease with cases of the first type, where different things are said to be the same in one respect only. But quantity raises a special problem, because it illustrates a case where 'two things may be both "the same" and "different" – *in the very same respect'*. The abstractionist can only construe this sort of difference as a combination of two independent abstraction processes. The absolutist theory, on the other hand, provides the means to at least acknowledge the difference between the two situations, and my guess is that this provides us with the reason why Russell chose to hold an absolutist conception of magnitude, even though he endorsed a relativist approach to cardinality. If this were right, then Levine would be wrong in considering the absolutist doctrine presented in PoM III as the remains of a Moorean doctrine which Russell no longer held in 1903. There was a strong reason (magnitudes have kinds, but numbers do not) not to deal with the two cases in the same way.

As I said, in many respects, this development is a digression which moves us away from our main line of interest. That I have nevertheless decided to include it here is because it provides us with another illustration of an important theme of this book: Russell, in his analyses, far from blindly applying certain general formal recipes, was very sensitive to certain fine-grained features of the subjects under discussion. Thus, the 'theorem of abstraction' allowed Russell to define magnitudes as sets of equal quantities. Yet, he refused to use it. Why? Because to apply to magnitudes the recipe which worked for cardinal numbers would have led him to pass over in silence one important difference between the two cases: the division of quantities into exclusive kinds. There is thus no incoherence in his choices. As in geometry, his fluctuations have their source in the willingness not to lose sight of the specificities of the conceptual landscapes he was considering.

4.4   The relational theory of distance

In section 4.2, we have seen that Russell split up the concept of extensive magnitude into two different subspecies: the magnitude of divisibility and distance. I will now come back to Russell's doctrine of distance. To expound it, I will refer to three sections of the English manuscript (1900b) of 'On the Logic of Relations' that he eliminated from the version he published in French in the *Rivista* one year later (1901a). I have several reasons for adopting this strategy. First, as I have said, the formal presentation Russell gave in 1900 is much clearer than the description he drew in 1903. Second, the connection to the works of other mathematicians (especially Burali-Forti) is made easier. Third, the 1900 version helps us to see the connection between the early theory of distance and the mature PM doctrine of the vector families.

Here is the formal definition of distance in Russell (1900b, § 7, p. 609):

7*1.1 $\Delta = FG \cap L \ni \{x, y \in \mathbf{D} . \supset_{x, y} . \ni L \cap R \ni (xRy) : Q = R_L .$
$R_1, R_2, R_3 \in L . R_1QR_2 . \supset_{R_1, R_2, R_3} . R_1R_2 = R_2R_1 . R_1R_3 \, Q \, R_2R_3\}$ Df

Let me first explain the notation: Δ designates a kind of measurable relational magnitude (i.e. a kind of distance), *F* is the class of dense series, *G* the set of groups. A distance is thus a densely ordered group (this is the meaning of 'Δ = *FG*...'), which is submitted to some further constraints (expressed by the rest of the formula). Nothing particularly surprising here: one recognizes the standard combination between an ordinal (dense series) and an additive structure (group) that is already found in Burali-Forti for instance. What is original, however, is the way Russell defined the concept of a group. How can we account, in the new relational framework, for the group operation? Is it a three-term relation, a combination of a relation with identity, or a new kind of term?

Russell set out the following definition (1900b, § 2, p. 594):

2*1.1   Group = *G* = Cls'1→1 ∩ *K* ∍{*P* ∈ *K* . ⊃ *P*⁻¹ ∈ *K* : *P*, *R* ∈ *K* ⊃_{P, R} .
          PR ∈ *K*. π = ρ} Df

A group is here described as a set *G* of one–one relations, such that:

1. If *P* is an element of *G*, *P*⁻¹, the converse of *P*, is an element of *G*.
2. If *P* and *R* are elements of *G*, the relative product of *P* and *R* belongs to *G*.
3. Every relation of *G* has the same domain (which I will call **D** in the following).[42]

The relative product *PR* of two binary relations *P* and *R* is the relation which holds between any terms *x* and *y* when there is a *z* such that *xPz* and *zRy*. Thus, 'the relative product of *brother* and *father* is *paternal uncle*; the relative product of *father* and *father* is *paternal grandfather*; and so on' (Russell and Whitehead 1910, p. 269). Note that the concept of relative product is more general than the standard notion of composition of mappings: any two binary relations (not only any two mappings) can be 'multiplied'.[43] However, in the definition 2*1.1, the relations considered are all mappings of **D** onto **D** (clause 3),[44] and the relative product can be regarded as a standard composition operation. Now, clause 1 says that every member of the group has an inverse with respect to the composition operation, while clause 2 requires that *G* is closed with respect to the same operation. Finally, one can verify that the relation identity (which transforms any element of **D** onto itself) belongs to *K* (since, if *P* is in *G*, *PP*⁻¹ is in *G*). To summarize: *G* is here defined as a set of one to one transformations of **D** onto itself, which forms a group with respect to the composition operation – or shorter: a group is here defined as a *transformation group*.

Today, it is standard to define a group in an 'abstract way', as a structure ⟨*S*, +⟩ (where *S* is a set and + is a binary operation) which satisfies the usual conditions.[45] A very important theorem of Cayley (1878), already well-known in Russell's time, showed that the notions of abstract group and transformation group are actually equivalent: to each group defined abstractly, one can correlate a group of transformations; and to each group of transformations, one can correlate an abstract group. Russell had then the right to define the concept of group in the way he did. But one could nevertheless consider that the abstract characterization is simpler than the other one. A transformation group is a three-levelled structure: one should distinguish the elements of **D**, the transformations defined on **D** (which are the elements of the group G), and the composition (the relative product) between the transformations.[46] In an abstract group, we find only two sorts of ingredients: the elements of the group and the group operation defined on them. So, why did Russell choose the more complicated characterization? The answer is quite easy: in the abstract definition, the group operation has to be introduced as a new indefinable, while, in the transformation approach, the group operation is a concept, the relational product, which belongs to the general logic of relations. From Russell's point of view, the transformation outlook is thus very natural, since it allows us to present the theory of groups as a branch of the theory of relations.

Let me come back now to the notion of a distance. According to the definition 7*1.1 quoted above, a distance Δ is a densely ordered transformation group defined on a set **D** such that:[47]

1. The group operation is commutative: if R_1 and R_2 belongs to L, then $R_1R_2 = R_2R_1$.
2. The relative product is compatible with the order: if Q is the order relation, then, if R_1QR_2 then $R_1R_3 \ Q \ R_2R_3$.
3. The action of Δ on **D** is 'transitive',[48] in the sense that if x and y are any two terms of **D**, then there is a relation R in Δ which relates them.

In the course of his development, Russell added two new conditions, the Archimedean postulate, according to which, 'given any two distances of a kind, there exists a finite integer n such that' the nth power of the first distance is greater than the second distance' (PoM, p. 254),[49] and another postulate, called the postulate of linearity (or Du Bois Reymond's postulate), according to which 'any distance has a nth root, where n is any integer' (ibid.).[50] Thus, Russell's 1900 notion of distance is a transitive action of a densely ordered Archimedian Abelian group of transformation on a set **D**.

As such, Russell's distance is very close to Burali-Forti's homogeneous magnitude (see pp. 108–9). In this way, the positive cone of a continuous Russellian distance satisfies Burali-Forti's postulates. But there is nevertheless a very important difference between the two approaches.[51] Burali-Forti characterized the notion of magnitude as an abstract algebraic structure, while Russell adopted the transformation perspective. For Burali-Forti, a homogeneous magnitude is any structure $\langle G, + \rangle$ satisfying the axioms.

G could be a set of relations and + could designate a relative product, though this was not implied by the definition of magnitudes. For Russell on the other hand, distances are relations, and an addition of distances is always a composition of transformations (relative product). This is the emphasis on the transformation point of view that marked out Russell from his contemporaries, and it is the same emphasis that one finds again in his description of distance in PoM 21 (see pp. 117–18). In fact, the fit between Russell (1900b) and PoM is not perfect, since what he defines in the latter corresponds to the positive cone of the structure he described in the former.[52] But he varied on this point in PoM.[53] And above all, what is brought forward in PoM is the idea that distance should be seen as an ordered transformation group (or semi-group), and not as an abstract algebraic structure. Let me quote Russell again, putting the relevant passages in italic (PoM, p. 253):

> Every distance is a one–one *relation*; and in respect to such *relations* it is convenient to respect the *converse of a relation* as its –1th power. Further the *relative product* of two distances of a kind must be a distance of the kind. When the two distances are mutually converse, their *product* will be identity, which is thus one among distances (their zero in fact), and must be the only one which is not asymmetrical.

In 1900 as in 1903, distances are characterized as relations, and addition of distances as the relative products between relations. This was Russell's key insight. What made his contribution unique was not the specific axioms he adopted, but the way he formulated them – that is, his use of the logic of relation to define the group structure in terms of transformation groups. Nobody except him[54] used the notion of transformation to set out a definition of extensive magnitude.

It remains to explain how the Russellian distance could be measured. Recall that this was the aim of Russell's construction – distances were introduced in PoM as one of the two subspecies of the genus 'measurable magnitude'. Let me briefly explain his reasoning. Let *U*, belonging to Δ, be the distance unity (that is, the distance to which one attributes the measure 1). To prove that distances are measurable, Russell could have shown that there is a unique one–one mapping **m** from Δ into $\mathbb{Q}$ such that $\mathbf{m}(U) = 1$, and, *R* and *S* being any two members of Δ, that $R<S$ implies $\mathbf{m}(R)<\mathbf{m}(S)$ and $\mathbf{m}(RS) = \mathbf{m}(R) + \mathbf{m}(S)$.[55] He did not proceed this way however, and chose instead to focus on a particular mapping, the 'logarithm' function, which associates to each power of the distance unity its exponent. It is easy to show that, *n* being any integer, the relation $P = U^n$ belongs to Δ, and that the function which associates to each U^n the integer *n*, is one–one and satisfies the two conditions $U^n > U^m \Rightarrow n>m$ and $U^n U^m = U^{n+m}$. This function can be considered as a measure of the powers of *U*.

However, Δ being a dense series, there are many elements in Δ which cannot be represented as a power of U. Russell (1900b, p. 610) then extends his notation in the following way:

7*3.5 $\forall n, m \in \mathbb{Q}, R' = R^{m/n}$ iff $R'^n = R^m$ Df

Thus, $U^{1/n}$ is the relation V such that $V^n = U$. This is not an original stipulation at all – the move can be traced back to Euclid's notion of (commensurable) ratio. Two segments R and R' have the ratio m/n iff n steps of length R' will give exactly the same length as m steps of length R. Even if Russell does not take the trouble to prove it, one can easily show that, whatever the rational numbers q one takes, U^q belongs to Δ. Furthermore, it could be proven that the 'logarithm' function is one–one and fulfils the two conditions $U^q > U^{q'} \Rightarrow q > q'$ and $U^q U^{q'} = U^{q+q'}$. Thus, this extension of the notation leads naturally to a consideration of the logarithm function as a measure of the rational powers of U. The case of irrational powers is more complicated. Russell's (1900b) account is too sketchy to be intelligible (see 7*4.1 , p. 611). I will come back to this issue in the next chapter, when dealing with the PM theory of measurement. Anyhow, the treatment of the rational case is sufficient for understanding how Russell proceeded: instead of seeking all the mappings from Δ to $\mathbb{Q}$ which preserve the ordinal and additive structures, he exhibits one of them (the logarithm function) and identifies it as the measure function (modulo the choice of the unit). Viewed from the contemporary measurement theory perspective, his argument is wanting – he gives us one possible measure function only and does not address the question of its uniqueness. But according to the standard of the time, he is quite careful – although he does not show that the logarithm function is unique, he does demonstrate that it preserves the relevant structure.[56]

Russell (1900b) does something else in § 7: he shows that the measure function of the distances induces a measure function on **D**. As he wrote in PoM (p. 181):[57]

> Numbers are also assigned by [the method of distance] to the members of the class between which the distances hold; these numbers have, in addition to the arbitrary factor, an arbitrary additive constant, depending upon the choice of origin.

Russell begun his construction by defining a relation K between couples of **D** and elements of Δ:

7*1.7 $(x, y)KR$ iff xRy Df

Owing to the transitive condition, K is defined on $\mathbf{D} \times \mathbf{D}$. But nothing Russell has said so far guarantees that K is many–one, and his reasoning is wanting here.[58] Let us pass over this problem and assume that the action of

Δ on **D** is not only transitive but faithful (that there is exactly one relation in Δ which relates any two elements of **D**). Russell then defined a relation of equality and order between couples of **D**:

$$(xy)=(zw) \text{ iff } (x, y)KK^{-1}(z, w) \qquad \text{Df}$$
$$(xy)>(zw) \text{ iff } (x, y)KQK^{-1}(z, w) \qquad \text{Df}$$

The couple (x, y) has to the distance R the same relation that a quantity has to its magnitude: (x, y) and (w, z) are different couples of **D×D**, but they can have the same magnitude if xRy and wRz. The new symbol (xy) designates thus the quantities associated with the relational distances. Relying on his exponent notation, Russell next introduces a new product operation. Let n be any integer, and x, y, w and z be any elements of **D**. He stipulates that:

7*3.13 $(zw)=n(xy) \text{ iff } \exists R \in L, xRy \wedge zR^{n}w$ Df

He continues by defining the rational multiple of (xy):

7*3.51 $(xz)=m/n(xy) \text{ iff } \exists R \in L, xRy \wedge xR^{m/n}z$ Df

When an origin, say x, and a unity, say (xy), is fixed, to every rational number r, there can be associated (xz), as a rational multiple r of (xy). One can then show that this mapping is one–one, and preserves the order and the additive relations – in brief, that r could be taken as a measure of the (xz). A final easy step allows us to fix once and for all the origin of the coordinate system in x and to take the rational number r as a measure of the element z of **D**. By the end of § 7 (1900b), Russell has shown (or, at least, has given his reader enough hints to understand) how to assign to each element of a dense subset of **D** (provided that a point origin and a distance unity have been chosen) a unique rational number, in such a way that the ordinal and the additive structures are preserved.[59]

One could thus summarize Russell's reasoning in this way. Russell started with one of the several axiomatic systems of extensive magnitude that blossomed at the time – most likely, Burali-Forti's. He then attempted to accommodate this system within the framework of the logic of relations. To fulfil this task, he redefined the notion of a group: rather than seeing it as an abstract structure, he characterized it in terms of transformations. This was the core insight of the theory, and this is an insight that will be used again in PM VI. Then, he showed how relational distances can be associated with numbers in a way that preserves the ordinal and additive relations between them. Finally he induced from this last move a measure of the elements to which distances applied. Contrary to what happened with magnitudes of divisibility, the theory of distance could really be considered as a viable theory of measurable extensive magnitude.

Before concluding, let me emphasize once again the connection between PoM III and PoM VI. We have seen in section 4.2 that the distinction between magnitude of divisibility and distance was adjusted to the distinction between the empirical notion of length and the projective concept of distance. One can see now that there is a link between Russell's doctrine and the issue concerning the introduction of coordinates on a line. Indeed in Russell (1900b), a rational coordinate system on **D** is derived from the structure 'distance'. This connection between relational distances and coordinate systems is explicitly made in PoM § 411 (pp. 427–8). The details of the argument are complicated, but the underlying idea is simple: instead of constructing the Möbius net by an iteration of the quadrilateral construction, Russell generated it through a repetition of a certain projective transformation.[60] The transformation played there exactly the same role as the unit distance U in Russell (1900b). In other words, the theory of relational distance constituted the general framework in which the more particular question of the coordination of projective line could be inserted. As we will see, this connection between the notion of relational magnitude and the issue of coordination will become tighter and more apparent in PM.[61]

4.5 Conclusion

As I said at the beginning of this chapter, PoM III is often considered in the literature as a peripheral mire which should be cautiously avoided in order not to lose the main thread of the story that Russell is telling. We can now see how unfair this assessment is. Russell's doctrine of quantity is very elaborate and ambitiously systematic. It was based on three pillars: the distinction between magnitudes and quantities, the division of the sphere of magnitudes into exclusive kinds, and the claim that certain magnitudes are relational. On this basis, he developed an account of extensive (i.e. measurable) magnitude. This was not an easy task, since the indivisibility of magnitudes prevented him from introducing a primitive operation of addition. He got around the obstacle by breaking up the genus of extensive magnitude into two subspecies: the magnitudes of divisibility and distance.

The theory does have some weak points. The theory of magnitude of divisibility notably was never fully worked out. But it also contains several gems. Russell seemed to have anticipated Johnson's distinction between determinable and determinate;[62] the idea of challenging the unity of extensive magnitude was a clever insight, which allowed Russell to distinguish and articulate two concepts of metrical space (the projective and the empirical); last but not least, the relational theory of distance is a very promising approach,[63] which will be extended and generalized in PM VI. In the end then, PoM III does not deserve at all the oblivion into which it has been taken.

Despite the real strength of Russell's account, it remains that the developments on quantity were in 1903 nothing but 'concessions to tradition'. This was a direct consequence of Russell's approval of Dedekind's and Cantor's theory of continuity.[64] Contrary to what was held by the mathematicians of the eighteenth century, and contrary to what contemporary neo-Kantians still thought (see PoM chapter 51 on Cohen), Russell considered that no reference to quantity was needed to account for the continuity of the real line. Thus, even if he resisted the arithmetization of geometry, in PoM he accepted the arithmetization of real analysis: the theory of real numbers did not depend, in any way, on quantitative notions. As we will see now, this changed in 1913. In PM, Russell and Whitehead opposed the arithmetical definition of real numbers – the two philosophers aimed at coming back to the old-fashioned algebra which connected rational and real numbers with quantities. The place of the theory of magnitude became then more central; but, for all that, the main outlines of the doctrine of magnitude remained the same. More precisely, Russell and Whitehead extended the theory of distance to elaborate a general quantitative account of rational and real numbers. It is to this treatment that I will now turn.

5
Quantity in *Principia Mathematica*

In the Introduction, I quoted the letter Whitehead wrote to Russell on 14 September 1909:

> The importance of quantity grows upon further considerations – <u>The modern arithmetization of mathematics is an entire mistake</u> – of course a useful mistake, as turning attention upon the right points. It amounts to confining the proofs to the particular arithmetic cases whose deduction from logical premisses forms the existence theorem. But this limitation of proof leaves the whole theory of applied mathematics (measurement, etc.) unproved. Whereas with a true theory of quantity, analysis starts from the general idea, and the arithmetic entities fall into their place as providing the existence theorems. To consider them as the sole entities involves in fact complicated ideas by involving all sorts of irrelevancies – In short the old fashioned algebras which talked of 'quantities' were right, if they had only known what 'quantities' were – which they did not.

Whitehead was here setting up the programme which would be developed in PM VI. The key idea of the whole construction was supplied by (what we will call in the following) the 'application constraint' (written hereafter as *Applic*), according to which the definition of rational and real numbers should explain how these numbers are used to measure quantities. As Whitehead noted in his letter, the arithmetization approach does not satisfy *Applic* – it 'leaves the whole theory of applied mathematics (measurement, etc.) unproved'. It was to remedy this defect that Russell and Whitehead attempted to defend another definition of numbers. My aim in this chapter is to show how they intended to fulfil this programme. I will not however investigate why it was that they thought *Applic* should be satisfied. This other issue will be the subject of the next chapter.

To my knowledge, no account of PM VI exists in the literature.[1] One reason for this neglect is certainly that the PM theory of quantity contains many complicated and esoteric notational devices (their meanings are explained in various parts of PM). In the following pages, I have tried to translate the basic symbolism and expound the main ideas in a more up-to-date terminology. Still to facilitate a reading of PM VI, I have also compared Russell and Whitehead's theory to three other accounts of real numbers: Dedekind's, Burali-Forti's and Frege's. This rapprochement will help us to bring forward the most important features of PM theory.

In the first section, I will briefly present the three definitions of numbers and quantity to which I compare Russell and Whitehead's account. Section 5.2 will be devoted to the PM theory of rational and real 'pure' numbers to be defended in Part A of PM VI. Section 5.3 will bear on the doctrine of vector families (magnitudes), expounded in Part B, and on the conception of measurement, developed in Part C of PM VI.[2] In section 5.4, I will come back to the comparison between Frege's doctrine of rational and real numbers and the theory of vector family presented in PM. Russell and Whitehead shared with Frege the idea that application should be built into the definition of numbers. But the way they develop their insight was very different from Frege. I will claim that PM theory achieves in a more efficient way the task they both intended to fulfil.

5.1   Dedekind, Burali-Forti and Frege on real numbers

The second half of the nineteenth century was a period during which mathematicians progressively abandoned the traditional definition of mathematics as a science of quantity. As Epples (2003) has explained:

> During the 18th and part of the 19th centuries, many scientists agreed with the idea that mathematics was the 'science of quantity'. This science was understood to consist of the geometric or algebraic study of numbers and continuous magnitudes such as lengths and weights or their 'abstract' counterparts. During the 19th century this image of mathematics changed profoundly, and one may reasonably call this change the end of the paradigm of the science of quantity.

Weierstrass's, Dedekind's and Cantor's arithmetical definitions of the real numbers field can be both considered as the crowning achievement of the new paradigm and as a point of no return. After their work, it was no longer possible to believe that continuous quantity cannot be reduced to discrete numbers, and a fundamental tenet of the old model, i.e. the distinction between two species of quantity, exploded. One must be careful, however, not to impose an overstated homogeneity upon the mathematical world of the end of the nineteenth century. Many mathematicians remained attached to the idea that mathematics was the science of quantity; and there was notable resistance to the new arithmetization programme. Here I will draw attention to two such critical reactions: Burali-Forti's and Frege's accounts of real numbers.[3]

But first, a brief word on Dedekind's notorious construction in 'Continuity and Irrational Numbers'. Dedekind began his treatise (section 1) with a presentation of the ordered field of rational numbers (as the notion of quotient structure is at the heart of his work on ideal theory, he could have very easily defined the rational number field as the 'quotient' of $\mathbb{Z} \times \mathbb{Z}$ by the equivalence relation ~, where $(m, n) \sim (p, q)$ if and only if $mq = np$). In sections 2 and 3, he drew a 'comparison of the rational numbers with the points of a straight line' to show that 'there are an infinite number of points which correspond to no rational number'. He then wrote (1872, p. 8):

> If now, as is our desire, we try to follow up arithmetically all phenomena in the straight line, the domain of rational numbers is insufficient and it becomes absolutely necessary that ... the creation of the rational numbers be essentially improved by the creation of new numbers such that the domain of numbers shall gain the same completeness, or as we may say at once, the same *continuity*, as the straight line.

In sections 4 to 7, he explained how to 'create' the new numbers. He first defined the cuts over the rational numbers. A cut is a partition of $\mathbb{Q}$ into two sets C_- and C_+ such that each element of C_- is smaller than each element of C_+. He then introduced an order within the set U of all the cuts, and showed that $\langle U, < \rangle$ is complete, in the sense that the operation consisting in forming new cuts over $\langle U, < \rangle$ does not 'enlarge' the structure. He then introduced an addition and a multiplication and proved that $\langle U, 0, 1, +, . \rangle$ is a complete ordered field. He then showed that the Bolzano-Weierstrass theorem (i.e. 'the theorem that every magnitude which grows continually, but not beyond limits, must certainly approach a limiting value' (ibid., p. 2)) is satisfied in the new structure. With this last proof, he illustrated how the differential calculus could be developed within a purely arithmetical setting.

It is well known that Dedekind did not identify the real numbers with the cuts over the rationals. His only claim was that there is an isomorphism between $\langle U, 0, 1, +, . \rangle$ and the real numbers field. It seems that, in PoM, Russell went one step further and identified the two structures. This statement must be qualified since, instead of using cuts, Russell, following Peano, grounded his own construction of reals on segments of the rationals.[4] A segment is defined as the first element of a Dedekindian cut (for instance, with the notation used above, C_- is a segment). But this is a detail. The essential point is that Russell claimed that, contrary to what Dedekind suggested, real numbers are nothing else than segments of rationals (PoM, pp. 279–80). Russell's worry was that if a distinction were made between segments (or cuts) and real numbers, then the existence of the latter would become dubious.[5] I will come back to the PoM theory of real numbers in Chapter 6.

As for now, what is important is that, in this arithmetical tradition, the real line and real analysis were developed in a purely arithmetical way, without any recourse to quantity.

I will now turn to another approach, which explicitly follows a course opposite to the one taken by Dedekind. In the introduction (1898, p. 34) Burali-Forti claimed:

> Chapter I of this book contains the properties of the magnitudes which do not depend on the idea of number (integer or fraction or irrational) ... Chapter II contains the basis of the theory of the whole numbers. The idea of a whole number is logically derived from the usual and concrete idea of magnitude ... An analogous procedure is followed in chapters III and IV, devoted to the rationals and the irrationals ... [In taking this path, it is possible] to obtain the general idea of number in a concrete shape by deriving it from the concrete usual idea of magnitude, which is essential for the metrical part of geometry as well.

In a complete reversal of the arithmetization programme, Burali-Forti attempted, in his treatise, to base arithmetic and real analysis on the concept of quantity. To do so, he developed a two-levelled approach: first, he developed an axiomatic theory of quantity; second, he introduced numbers as ratios between quantities. In Chapter 4, I presented Burali-Forti's characterization of homogeneous magnitude $\langle G, + \rangle$ (see pp. 108–9).[6] From this basis, he undertook to define the different sorts of (positive) numbers. I will skip here his considerations about whole numbers to come to the introduction of positive rationals and reals. A rational is defined as an equivalence class over a subset of $G \times G^*$: two couples of magnitudes (q_1, q_2) and (q_3, q_4) belong to the same 'rational' equivalence class (define the same positive rational number) if and only if there are two positive integers m, n, such that $nq_2 = mq_1$ & $nq_3 = mq_4$. There is nothing surprising in this definition (since it maintained Euclid's characterization of Euclidean ratio); and showing that the rational numbers, so characterized, have all their expected properties, is an easy task (once it is admitted that there is a model of homogeneous magnitude).

How did Burali-Forti define real numbers? Real numbers are still regarded as relations between homogeneous magnitudes – but the definition is more complicated than in the rational case. Two couples of magnitudes (q_1, q_2) and (q_3, q_4) belong to the same 'real' equivalence class (define the same real numbers) if and only if there is a 'Russellian' segment θ of rational numbers (on this notion, see p. 136) such that $\lim \theta q_1 = q_3$ & $\lim \theta q_2 = q_4$, where $\lim \theta q_1 = q_2$ means that:

1. for all rational $r \in \theta$, $rq_1 \leq q_2$, and
2. for all rational $r \notin \theta$, $rq_1 \geq q_2$

This characterization is equivalent to Euclid's definition one finds in the *Elements*.[7] The new equivalence relation ($\lim \theta q_1 = q_3$ & $\lim \theta q_2 = q_4$) 'extends' the one used in the definition of rationals, and one can easily check that the real numbers, so defined, have all their expected features (at least, when one accepts that there are homogeneous magnitudes).

Burali-Forti's theory is thus a return to the old Euclidean algebra of ratios. The only difference is that the quantitative structure is here completely formalized. In particular, Burali-Forti used Dedekind's work for characterizing the continuity of the homogeneous magnitude and to define the real ratio. But still, despite his use of Dedekind's notions, the guiding line of Burali-Forti's theory stands in sharp contrast to the arithmetized approach: instead of placing the natural numbers at the centre of the mathematical edifice, Burali-Forti promotes the idea that rational and real numbers are nothing else than ratios of quantities. Magnitude thus appears to be more fundamental than number. This approach has an advantage over its rival. Since rational and real numbers are just ratios of quantities (modulo a certain given unit), the issue of application of numbers in measurement finds a clear and neat solution in Burali-Forti – which is not the case, in Dedekind, who severed the reals from any connection to quantity. But the satisfaction of *Applic* is balanced by an obvious weakness: in Burali-Forti's construction, the fact that the rational and real numbers have all their usual properties is dependent upon the existence of a homogeneous magnitude. Now, it might be that there are not enough quantities in the real world to satisfy the postulates of ordinal density or of ordinal completeness of Burali-Forti's system.[8] If this were the case, then the rational and real numbers, defined as ratios, would lose their arithmetical properties. In other words, in Burali-Forti, the fact that the numbers have the properties they have is grounded on the physical structure of reality.[9] This is precisely the view that Dedekind and Russell sought to avoid.

Let me come now to Frege's standpoint, developed in the second volume of *Grundgesetze* (1903b, § 153). This conception can be seen as an attempt to reconcile the two opposite standpoints of Dedekind and Burali-Forti:

> The path to be pursued here thus lies between the old way of founding the theory of irrational numbers, the one H. Hankel [or Burali-Forti] used to prefer, and the paths followed more recently [by Cantor, Weierstrass and Dedekind]. We retain the former conception of real number as a relation of quantities ... but dissociate it from geometrical or any other specific kinds of quantities and thereby approach more recent efforts. At the same time, on the other hand, we avoid the drawback showing up in the latter approaches, namely that any relation to measurement is either completely ignored or patched on solely from the outside without any internal connection grounded in the nature of the number itself ... Our

hope is thus neither to lose our grip on the applicability of arithmetic in specific areas of knowledge nor to contaminate it with the objects, concepts, and relations taken from those areas and so to threaten its peculiar nature and independence.

Frege wanted to retain, from the traditional Euclidian conception, the idea that the real numbers should be construed as relations between quantities. But for all that he did not want to renounce the view that the properties of real numbers are deducible from logic alone. In other words, Frege was looking for a mid-way solution, distinct from those proposed by both Dedekind and Burali-Forti.

Recently, Frege's approach has been continued by Hale (2000, 2002).[10] As the differences between Frege's own doctrine and Hale's theory are not relevant for my project, I will here focus on the latter, which is simpler and closer to Burali-Forti and Dedekind than the former. As with Burali-Forti, Hale introduces the numbers as relations between quantities. More precisely, he distinguishes three (successively richer) quantitative domains (henceforth, q-domains):

1. A normal q-domain is the positive cone of a totally ordered Archimedean Abelian group $\langle X, \angle, \oplus \rangle$;
2. A full q-domain $\langle X_f, \angle, \oplus \rangle$ is a normal q-domain which is densely ordered;
3. A complete q-domain $\langle X_c, \angle, \oplus \rangle$ is a full domain whose every bounded-above non-empty $S \subset X_c$ has a least upper bound.

As an example of a normal q-domain, one can give the positive integers; as an instance of a full q-domain, one can give the positive rational numbers; and as an example of a complete q-domain, one can think of the positive real numbers.

After having presented an axiomatic of a quantitative structure, Hale sets out a definition of number, which conveys the gist of Euclid's notorious abstractionist definition of ratio: 'ratios $a : b$ and $c : d$ are the same just if equimultiples of their numerators stand in the same order relations to equimultiples of their denominators' (2000, p. 107). In symbolic notation:

$$\forall a, b, c, d \in X \, [a : b = c : d \Leftrightarrow \forall m, n \in \mathbb{N} \, ((ma \angle nb \Leftrightarrow mc \angle nd) \vee$$
$$(ma = nb \Leftrightarrow mc = nd) \vee (nb \angle ma \Leftrightarrow nd \angle mc))]$$

When the q-domain X_f is full, the following principle (called 'the principle of the fourth proportional' in the Euclidean tradition) holds:

$$\forall a, b, c \in X_f \, \exists q, q' \in X_f (a : b = c : q \, \& \, a : b = q' : c)$$

This allows Hale to introduce order, addition and product on the set of ratios in the following way:

$a : b < c : d$ iff $\exists q' \, (a \angle q \,\&\, c : d = q : b)$
$a : b + c : d = (a + q') : b,$ where q' is such that $c : d = q' : b$
$a : b \times c : d = a : q$ where q is such that $c : d = b : q$

A rational is then defined as a ratio between elements of a full q-domain, and a real as a ratio between elements of a complete q-domain. It is then an easy task to show that the numbers, defined as they are, have all their expected arithmetical properties. Note that here, as in Burali-Forti's case, the algebraic and order theoretic characteristics of ratios are inherited from the algebraic and order theoretic features of their respective q-domains. It is only because the full (complete) domain is defined as densely (completely) ordered that the ratios are densely (completely) ordered.[11]

Up to here, Hale's theory is very close to Burali-Forti's: both philosophers adopt a two-levelled account, whose first part is constituted by an axiomatic theory of quantity, and whose second part is devoted to the definition of number as ratio. They both prove that, 'provided there exists at least one complete q-domain [one homogeneous magnitude], we can introduce the positive real numbers, by abstraction, as the ratios on that domain' (Hale 2000, p. 108). Burali-Forti's and Hale's Fregean paths begin to diverge when they come to the issue of whether the conditional 'provided there exists at least one complete q-domain' could be logically satisfied. We have seen that, in Burali-Forti's construction, the existence of a complete q-domain remained an empirical matter that no a priori investigation can settle. On the contrary, Frege and Hale considered it to be of the utmost importance to prove the existence of a complete q-domain. If there were no such model, then the usual 'arithmetical' laws, which govern the real number field and, hence, the principles which form the basis of the differential calculus (like the Bolzano-Weierstrass theorem), will not be true. For a logicist, such a situation is intolerable.

Hale and Frege thus add to the two levels previously distinguished (q-domains, ratios) a third conceptual layer, intended to provide us with the missing existence theorem. Hale begins by showing that the set of natural numbers (abstractly defined by Hume's principle) is a normal q-domain, in the sense that it satisfies the postulates characterizing the structure. After having defined $R^{\mathbb{N}}$, the set of ratios on $\mathbb{N}$, by using the standard construction, he then remarks that $R^{\mathbb{N}}$ constitutes a full domain (2000, p. 111). However, to get what is required (the completeness of the q-domain), a much stronger abstraction principle is needed. Hale introduces then the notion of 'cut property': 'informally, a cut-property is a *non-empty* property whose extension is a *proper* subset of $R^{\mathbb{N}}$ and which is *downwards closed* [i.e. $\forall a \forall b$ $(Fa \Rightarrow (b < a \Rightarrow Fb))$] and has *no greatest instance* [i.e. $\forall a(Fa \Rightarrow \exists b(b > a \Rightarrow Fb))$]'

(2000, p. 112). In other words, the extension of a cut property is a segment without a greatest element. At this stage, Hale defines a new abstraction principle, called '(Cut)', which allows him to introduce the object common to all coextensive cut properties. He then verifies that the set of these objects (the set of cuts), endowed with a suitable order and addition operation, forms a complete q-domain. This amounts to proving that there is at least one complete q-domain and salvages the quantitative definition of numbers from the physicalist disaster.

Hale's existence proof, then, is a sort of abstractionist rewriting of Dedekind's arithmetical construction of the reals. The idea is to use the cut construction, not, as was the case with Dedekind, in order to define real numbers, but in order to provide the existence theorems missing in the Burali-Forti-type approaches. For Hale, reals are ratios between elements of a complete q-domain; they are not cuts; but cuts can be nevertheless considered as elements of a complete q-domain. Frege did not adopt exactly the same method as Hale, but his strategy was essentially the same: referring to Weierstrass, instead of Dedekind, Frege provided an instance of a complete q-domain by relying on the construction of the arithmetizers.[12]

One can thus understand the sense in which the path pursued by Frege and Hale lies between the old way of founding the theory of irrational numbers taken up by Burali-Forti and the path followed by Dedekind. With Burali-Forti and contrary to Dedekind, Frege and Hale adheres to *Applic*, according to which the definition of the real numbers must account for its applicability to measurement. However, with Dedekind and contrary to Burali-Forti, Frege and Hale refuse to endorse a physicalist conception of the real number field. To avoid this, they take up the construction of the 'arithmetizers' and logically prove the existence of a complete q-domain. This subtle position is close in spirit to the one taken by Russell and Whitehead in PM VI. The two philosophers rejected both Burali-Forti's and Dedekind's doctrine for roughly the same reason as Hale and Frege. However, the way Russell and Whitehead developed their insight is completely different from the way Hale and Frege did. Before comparing the two doctrines, I will first present the PM theory of numbers and quantities.

5.2 Russell and Whitehead's theory of numbers

At the beginning of PM VI, Russell and Whitehead claimed that the introduction of negative, rational and real numbers (what they call generalization of numbers) is necessary as soon as one wants to account for measurement (1913, p. 233): 'the purpose of this Part is to explain the kinds of applications of numbers which may be called *measurement*, [and] for this purpose, we have first to consider generalizations of number'. They did not only mean that rational and real numbers are necessary for measurement (which is trivial), but that the very notion of rational and real numbers are tied up with quantities and measurement. That is, like Frege and Burali-Forti, Russell and Whitehead endorsed *Applic* and rejected, for this reason, the Dedekindian doctrine of real numbers.

There is, however, from the very beginning, a difference between PM and Burali-Forti's and Frege-Hale's construction. Section A of PM VI is devoted to numbers, section B to the vector family (PM's word for 'quantity') and section C to measurement – that is, unlike the other anti-arithmetizers, Russell and Whitehead did not begin their presentation with the theory of quantity. Why did they choose to start with numbers, instead of quantities? This issue is related to the distinction they made between (what they called) 'pure numbers' and (what they called) 'applied numbers'. At the beginning of section C, they explained (1913, p. 407):

> In this Section, the 'pure' theory of ratios and real numbers developed in Section A is applied to vector-families. A vector-family, if it has suitable properties, may be regarded as a kind of magnitude. In order to derive from the 'pure' theory of ratio a theory of measurement having the properties which we should expect, it is necessary to confine ourselves to some one vector-family; that is, instead of considering the general relation X, where X is a ratio, we consider the relation X [restricted to] κ where κ is the vector-family.

Roughly said, applied numbers in PM correspond to the numbers viewed as ratios of quantities, whereas pure numbers in PM have no equivalent in Frege-Hale theory. To understand the real import of Russell and Whitehead's notion of pure numbers, and to understand the distinction between their approach and the Frege-Hale view, there is no other way than to enter into more detail in PM VI.

Let me begin with the definition of pure rational number. In *120 of PM, Russell and Whitehead defined the inductive cardinals and the operations of addition and multiplication between them. In *302, they used these notions to define the relation Prm between two couples of integers (m, n) and (p, q):

$$(m, n)\mathrm{Prm}(p, q) =_{\mathrm{def}} m \text{ is prime relative to } n \ \& \ m \times q = n \times p$$

One can recognize, in the second part of the definiens, the equivalence relation used in the standard construction of $\mathbb{Q}^+$ from $\mathbb{N}$.

The ratio μ/v, with μ and v two positive integers ($v \neq 0$), was then defined in the following way (1913, p. 260):[13]

*303.01 $$\mu/v =_{\mathrm{def}} \hat{R}\hat{S}\{\exists(\rho, \sigma) \cdot (\rho, \sigma)\mathrm{Prm}(\mu, v) \cdot_{\exists} !R^{\sigma} \cap S^{\rho}\}$$

Two relations R and S have the ratio μ/v (written $R(\mu/v)S$) if and only if there are two objects x and y such that, ρ/σ being the irreducible[14] fraction associated with μ/v, both the relations R^σ and S^ρ hold between them. The symbol 'R^σ' designates here the relational product of R, σ multiplied by itself.[15]

One must distinguish two elements in the definition *303.01: the first one is the clause '$\exists(\rho, \sigma) . (\rho, \sigma)\text{Prm}(\mu, v)$'; the second one is the remainder of the formula '$\hat{R}\hat{S}\{\exists(\rho, \sigma) . {}_{\exists}!R^\sigma\cap S^\rho\}$', which states that a ratio is a relation between relations. These two ingredients do not play the same role. The former clause is used for giving the ratio all its standard arithmetical properties. Indeed, the definitions of addition, of multiplication and of order between ratios are based on it (1913, pp. 279, 284, 292):

$$m/n < p/q = {}_{\text{def}} m \times q < n \times p \qquad\qquad *304.01$$
$$(m/n) \times (p/q) = {}_{\text{def}} (m \times p)/(n \times q) \qquad\qquad *305.01$$
$$m/n + p/q = {}_{\text{def}} (mq + pn)/nq \qquad\qquad *306.41$$

With the help of the axiom of infinity, Russell and Whitehead showed that the set of ratios, equipped with these operations and with the ordinal relation, is a dense Archimedean ordered field.[16] All the expected arithmetical properties of rational numbers follow then from these definitions (and from the axiom of infinity). Mathematically speaking, what is essential in *303.01 is the relation Prm: it alone supplies the means to construct $\mathbb{Q}^+$ from $\mathbb{N}\times\mathbb{N}$. The second relational ingredient does not play any role in the derivation of the arithmetical properties of ratios.[17]

But if the relational part of *303.01 is mathematically empty, why did Russell and Whitehead include it in their definition? Why did they not follow custom and introduce the rationals as equivalence classes on $\mathbb{N}\times\mathbb{N}$? In a letter to Jourdain dated 21 March 1910, Russell, after having recognized that one could define a ratio as a relation between integers, added (Grattan-Guinness 1977, p. 130):

> I have now accepted from Whitehead a new quantitative (non-arithmetical) definition of μ/v, according to which it is a relation of vectors R, S which holds (broadly) whenever ${}_{\exists}!R^v \cap S^\mu$. This enables you to take two-thirds of a pound of butter without an elaborate arithmetical detour.

The function of the relational 'appendage' is then to explain the application of ratios to measurement, and this is in line with Russell and Whitehead's introductory statement according to which the generalization of numbers should be tied up to the operation of measuring quantities. The idea appears natural, once put in the context of Russell's 1900 and 1903 theory of distance (see Chapter 4). Recall that distances were viewed as a set of relations satisfying certain conditions (they should be one–one relations defined on the same field, belonging to a dense Abelian ordered group), and that Russell's aim was to show that, a given unit being chosen, one could define an isomorphism between $\langle \mathbb{Q}, <, + \rangle$ and ratios between distances. In this context, one could thus easily define the rational numbers as a relation between relation-distances, exactly as Euclid did. In fact, the condition $R(\mu/\nu)S$ iff there are two objects x and y such that $xR^{\nu}y$ and $xS^{\mu}y$ is close to the standard characterization of a commensurable ratio – two magnitudes R and S have the ratio (μ/ν) if ν steps of size R from a point x leads to exactly the same point as μ steps of size S.

One must be careful not to go too far in that direction, however. A kind of distance (as a complete q-domain in the Frege-Hale framework) was a very special kind of relational structure. Ratios, as they are defined in PM, can be applied to distances, but they can also be applied to relations which are not distance – they can be applied to any relation. This point is crucial. According to *303.01, R and S have the ratio μ/ν if and only if the intersection of R^{ν} and S^{μ} is not empty. Now, if no special conditions are set out, two relations R and S could have the ratio μ/ν without $R^{\nu} = S^{\mu}$. The structure distance is so regular that, if R and S are two distances such that $_{\exists}! R^{\nu} \cap S^{\mu}$, then R^{ν} and S^{μ} are the same. But Russell and Whitehead made it clear that this does not hold in general. Thus, two relations can have many distinct ratios (one can have $\mu \times \sigma \neq \rho \times \nu$, $_{\exists}! R^{\nu} \cap S^{\mu}$ and $_{\exists}! R^{\sigma} \cap S^{\nu}$) – which is impossible if the two relations were two distances of the same kind.[18] Furthermore a relation can have the same ratio with many distinct relations (one can have $_{\exists}! R^{\nu} \cap S^{\mu}$ and $_{\exists}! R^{\nu} \cap Q^{\mu}$, with $S \neq Q$); which, once again, is prohibited in structure distance. In other words, rational numbers, as defined in *303.01, cannot be regarded as measures of the relations they are applied to: one relation can have many different ratios with the unit, and a same ratio could relate the unit to many distinct relations.

The gap between the two conditions $_{\exists}! R^{\nu} \cap Q^{\mu}$ and $R^{\nu} = S^{\mu}$ is crucial since it is at the basis of the distinction between pure and applied ratio. As Russell and Whitehead remarked (ibid., p. 262): 'in practical applications ... when R and S are confined to one vector-family, different ratios do become incompatible, as will be proved at the beginning of section C'. Thus, the restriction of the field of ratios made in section C guarantees the equivalence between the two conditions – and guarantees as well that pure ratios can be regarded as measures or as applied ratios. But the crucial point is that, in section A, such a restriction is not made (ibid.):

So long as we are not concerned with the applications which constitute measurement, the important thing about our definition of ratio is that it should yield the usual arithmetical properties, in particular the fundamental property

$$\mu/\nu = \rho/\sigma \Leftrightarrow \mu \times \sigma = \rho \times \nu$$

Which is proved, with our definition, in *303.39. Thus any further restriction in the definition would constitute an unnecessary complication.

The definition *303.01, without any restriction with respect to the field of the relation, discloses the nature of 'pure' ratios, and pure ratios are what are needed to do arithmetic. When the field of ratios is restricted in a suitable way, to generate a structure akin to 1900/1903 distances (a measurable vector-family in the new jargon), then ratios will become applied ratios and can be viewed as measures of magnitudes. But the additional conditions are not presupposed in section A. All the arithmetic of rational numbers follows merely from the non-relational part of *303.01 and the logical Pp – there is no need to set additional constraints on the structure of the entities the ratios apply to.

The same distinction between a pure and an applied level occurred in the PM construction of real number. One actually finds two definitions of reals in section A. Real numbers were first introduced in *310 as segments of ratios. Russell and Whitehead maintained there the method used in PoM (to which they refer in PM 1913, p. 316). This definition is formally irreproachable, but the link with application is not clearly depicted. To remedy this defect, Russell and Whitehead, in *314, slightly amended their first characterization. Let S be a segment of ratios. Being a set of ratios (that is, a set which contains sets of couples of relations), S has not the same logical type as a ratio. Now, $\cup S$, the union of the sets of couples of relations contained in S, is a set of couples of relations; it has then the same logical type as a ratio. On the other hand, it is clear that to each segment S corresponds one and only one $\cup S$. Furthermore, Russell and Whitehead showed in *310.33 that the order between the segments (between the Ss) is similar to the order naturally induced on the unions (on the $\cup S$ s), and that one can, in the same natural way, introduce an addition and a multiplication on the class of all the $\cup S$. In other words, they proved that the replacement of S by $\cup S$ does not change anything to the arithmetical construction of the reals; the ordered field structure obtained in this way retains all the expected ordinal and algebraic properties. This second definition has however an advantage over the first one: a real number is now defined as a relation between relations, and no longer as a set of relations. Henceforth, in *314, real numbers have the same logical type as ratios – they can be directly applied to that which the ratios are applied.[19]

One finds then in the real case the same pattern of reasoning that one encountered in the rational case. Russell and Whitehead wanted to secure a definition of real numbers which gave them all the arithmetical properties they need in one shot; but at the same time, they were willing to satisfy *Applic*, and for this reason, they sought to characterize a real number as a relation between relations. Thus, as in the rational case, the definition set out in section A is not fully intelligible without taking into account the content of section C concerning measurement.

There is however a difference between the two situations. If it is easy to relate *303.01 to Euclid's definition of ratio, it is more difficult to make a connection between *314 and the measurement of magnitudes. In the

Frege-Hale framework, a real number is a ratio of two elements of a complete q-domain – that is, if a unit quantity is chosen, then to any quantity in the domain, a unique real measure will be associated. Now, this can never happen in Russell and Whitehead's account. From *314, it indeed follows that, for any two real numbers X and Y such that $X \leq Y$, $U(X)S$ implies $U(Y)S$.[20] If U is taken as a unit quantity, this means that if a certain relation S has the real 'measure' X, then it would also have all the real 'measures' greater than X. As Quine noted (1963, pp. 129–30): 'Whitehead and Russell's ratios were mutually exclusive ... whereas their real numbers were nested.' This fact does not forbid any connection between real numbers and quantities, but it reveals that the notion of measurement is not the same in the real case and in the rational case. I will come back to this point at the end of the next section.

5.3   The relational theory of vector family

In order to define ratios as measures of relations, Russell and Whitehead turned their attention to certain sets of relations called vector-families. The idea was that, when restricted to certain species of vector-family, pure ratios will become applied ratios, that is, measures of the elements (vectors) of the family. PM's vector-family corresponds to Frege's and Hale's q-domains. But, in section B, Russell and Whitehead elaborate their theory of magnitude in a very general way, without drawing any connection to the theory of numbers developed in section A. How then did they characterize the notion of a vector-family?

Russell and Whitehead first defined (*330.01) the concept of a correspondence over a set α, written $cr'\alpha$. A correspondence over α is the complete semigroup of injective mappings on α; or in other words, the set of one–one relations, whose domain is α and whose codomain is included in α, endowed with the operation of relative product (or composition).[21] Now, a vector-family κ defined on α is introduced (*330.03) by three conditions:

1. $\kappa \subseteq cr'\alpha$.
2. $\kappa \neq \varnothing$.
3. $\forall R, S \in \kappa,\ R|S = S|R$, where $R|S$ is the relative product of R and S.

A vector family of α is thus any non-empty commutative subset of one–one relations defined on α.

It is essential to realize that, compared to the Frege-Hale notion of the complete q-domain, or to Russell's former concept of a 'kind of distance', κ is a very weak structure. It has lost nearly all its algebraic content. κ is not necessarily closed under the composition operation, it is not necessarily a symmetrical structure, and the identity mapping does not necessarily belong to it. Among the algebraic properties, only the commutative condition remains. Regarding order, the situation is even worse: κ has no ordinal structure.

One thing, however, had stood the test of time: quantities (vectors) were still regarded in 1913 as forming a relational structure which operates over a set. That is, one finds again in PM VI the three-levelled schema first presented in Russell (1900b). One thus has (compare with pp. 112, 127–8):

Level 2: The relations between κ-vectors.
Level 1: The κ-vectors R, S, ...
Level 0: The entities belonging to α to which the κ-vectors apply.

Instead of speaking of a one–one mapping onto α, Russell and Whitehead considered in 1913 a commutative subset of one–one mappings onto α. And instead of postulating a very rich structure at level 2, they reduced the properties satisfied by the κ-vectors to a bare minimum, only requiring that the product be commutative. But despite these differences, PM vector families were, like PoM distances, defined as relational structures acting on a set – and the product of relations were in both cases viewed as the correlate of the quantitative addition. They were nothing but a generalization of the notion of a transformation group, which occupied in Russell (1900b) the centre of the scene.[22]

In the remainder of section B, the authors progressively enriched the concept of vector-family by setting out further constraints. I will here give a brief overview of the main directions that guided Russell and Whitehead's developments.

The connected family (*331) is the most fundamental species of family. A point a of α is said to be connected in κ if and only if any point of α can be reached from a through a κ-vector.[23] A connected family is a family which contains at least one connected element (*331.02). When every point is connected, the family is said to have connexity (*334.27). Connected families have very important properties. Firstly, in a connected family, κ operates faithfully on α; i.e. x being any point of α, R and S any two κ-vectors, the following holds:

$$R'x = S'x \text{ iff } R = S \text{ (*331.42)}$$

Secondly, when κ is connected, the relational product of two κ-vectors or its converse belongs to κ; furthermore, if R is a κ-vector, then R^n (n being any natural) or its converse belongs to κ. Connected families have thus a strong closure property. This led Russell and Whitehead to study in detail κ_l, 'the class of such relations as $R^{-1}S$, where R, $S \in \kappa'$ (1913, p. 342). One important result (*331.43) is that, when κ is connected, κ_l operates faithfully and transitively on α (i.e. in this case, there is one and only one element of κ_l which relates any two points of α).[24] I have just said that if R belongs to κ, then $S = R^n$ belongs to κ_l. But if R belongs to κ_l, then S is not in general a member of κ_l. Russell and Whitehead showed however that there is at most one element H in κ_l, such that $S \subseteq H$. They called H the representative of S. The importance of this notion comes from the theory of applied ratio. Indeed, in a connected family κ, if two relations L and M belong to κ_l and if $L(m/n)M$, then the representative of L^n (if it exists) is the same as the representative of M^m (if it exists).

After having dealt at length with connected families, Russell and Whitehead introduced a new kind of family, the open vector-family. κ is open (*333.02) if and only if $\forall x \in \alpha, \forall n \in \mathbb{N}, \forall S \in \kappa_l, S \neq Id \Rightarrow S^n x \neq x$ (*Id* designates here the restriction of the identity mapping to α). The idea behind this is simple: κ is open if and only if the successive applications of any κ_l-vector to any α-point never make us come back to our initial position. It is easy to show that the existence of an open family implies the axiom of infinity (*333.19).[25] Russell and Whitehead proved as well (*333.53) that, in an open connected family κ, if L and M are any two κ-vectors (both distinct from *Id*), and if $L(m/n)M$ and $L(p/q)M$, then $m/n = p/q$. This result is crucial for the theory of applied ratio, since it amounts to saying that two κ-vectors cannot be related by distinct ratios.

Russell and Whitehead then introduced the notion of a serial family (*334). Let x and y be any two elements of α. Then the relation $s'\kappa$ is defined in the following way: $x(s'\kappa)y$ iff there is at least one κ-relation R ($\neq Id$) such that xRy. Now, κ is said to be serial when $s'\kappa$ is a connected asymmetrical transitive relation – that is, a total order.[26] A key concept in the theory of the serial family is that of the transitive point. An α-point a is transitive if and only if 'any point which can be reached from a by two successive non-zero steps can also be reached by one non-zero step' (1913, p. 383).[27] A transitive family is a family which contains at least one transitive point. Russell and Whitehead showed that a transitive family that has connexity is serial (*334.03). They also proved that a connected family is transitive if and only if it is closed under the relational product (1913, pp. 383–4).

In serial families, it is α (and not κ) which is ordered. To introduce an order on κ, Russell and Whitehead defined (*336.01) V_κ as the relation that holds between two κ_l-relations R and S only in the case where there is an x in α such that $(R'x)(s'\kappa)(S'x)$. The restriction of V_κ to κ is the relation U_κ. They succeeded in characterizing the cases where V_κ and U_κ are ordinally similar to $s'\kappa$. They showed that if κ is connected and serial, V_κ defines an order on κ_l. They also proved that if κ is connected and 'initial' (i.e. that there is a connected α-point that cannot be reached by any non-zero κ-vector), then U_κ and $s'\kappa$ are similar. Finally, they showed that the relation U_κ is compatible with the relational product, in the sense that (under the assumption that κ is connected):

$$PU_\kappa Q \Leftrightarrow \exists T \ (\neq Id) \in \kappa, P = TQ \qquad \text{(*336.41)}$$
$$PU_\kappa Q \Rightarrow \forall R \in \kappa, (PR)U_\kappa(QR) \qquad \text{(*336.411)}$$
$$\forall n \in \mathbb{N}, PU_\kappa Q \Rightarrow P^n U_\kappa Q^n \qquad \text{(*336.511)}$$

Russell and Whitehead then introduced the notion of the submultipliable family: κ is submultipliable (*351.01) when and only when $\forall R \in \kappa$, $\forall n \in \mathbb{N}$, $\exists S \in \kappa$, $R = S^n$ (in Russell (1900b), this is the content of the Du Bois Reymond's postulate). They showed that if the converse of the relation $s'\kappa$ is serial, dense and semi-Dedekindian,[28] then κ is submultipliable.

Let me stand back from the detail and summarize the reasoning. After having defined the very weak concept of vector-family, Russell and Whitehead introduced the central notion of connected family. They immediately restricted again the scope of their inquiry by focusing on the open connected family. At this stage, their main difficulty was the lack of closure of open connected families. To get round the obstacle, they extended κ into κ_l and developed the theory of representatives. They succeeded to show that any two vectors belonging to an open connected family have at most one ratio. At this point, they introduced a relation of order. They first proved that, if κ is transitive and has connexity, $s'\kappa$ defines an order on α; also that, if κ is serial and initial, an order can be induced from α onto κ itself. The relation of compatibility between the so-defined order and the relational product was discussed. Finally, the concept of submultipliable family was introduced. Of course, this survey is far too sketchy to do justice to the richness of the developments contained in section B of PM VI.[29] But despite its incompleteness, this panorama does not leave any doubt about Russell and Whitehead's intent. Starting from a very weak and general structure, the two philosophers sought to recover a configuration akin to the one Russell (1900b) described under the name of 'distance'. The final stage of the construction is indeed represented by the open connected serial submultipliable family κ – a structure which is akin to the positive cone of distance.[30] Russell and Whitehead wanted then to show how the rational and real numbers of section A, when restricted to a certain sort of vector-family, could be viewed as measures of the vectors of the family. Thus, the entire development of section B aimed at supplying the resources to elaborate the theory of 'measurable family' and 'applied ratio' expounded in section C.

Let me briefly explain the main components of this last doctrine. A measurable family is defined at the beginning of section C by the following four conditions (1913, p. 407):

(1) No two members of a family can have two different ratios ...
(2) All ratios [except 0 and ∞] must be one–one relations when limited to a single family ...
(3) The relative product of two applied ratios ought to be equal to the arithmetical product of the corresponding pure ratios with its field limited, i.e. if X, Y are ratios, we ought to have

$$X \updownarrow \kappa \mid Y \updownarrow \kappa = (X \times Y) \updownarrow \kappa \text{ or } X \updownarrow \kappa_l \mid Y \updownarrow \kappa_l = (X \times Y) \updownarrow \kappa_l.$$

That is to say 'two-thirds of half a pound of cheese' ought to be $(2/3 \times 1/2)$ of a pound of cheese; and similarly in any other case ...

(4) If X, Y are ratios, and T is a member of the family κ, we ought to have

$$(X \mathbin{\updownarrow} \kappa'T) \mid (Y \mathbin{\updownarrow} \kappa'T) = (X+Y) \mathbin{\updownarrow} \kappa'T,$$

that is two-thirds of a pound of cheese together with half a pound of cheese ought to be $(2/3 + 1/2)$ of a pound of cheese, and similarly in any other instance.

It will be helpful to translate these clauses into a more palatable notation. Let $\mathbf{M}_S(R)$ be the relation which associates to each κ-vector R, the ratios, if they exist, that link R with S, S being a κ-vector arbitrarily chosen (the 'unit'). Condition (1) amounts to saying that the relation $\mathbf{M}_S(R)$ is many–one: there is at most one ratio which is associated with any κ-vector R (S being the unit). Condition (2) is very strong: it requires that, whatever the (strictly) positive rational r one considers, there is a unique vector R such that $\mathbf{M}_S(R) = r$. The first two conditions require that, a unit being chosen, there is one and only one one–one mapping from $\mathbb{Q}^+$ onto[31] κ. Recall that these conditions were not satisfied in the general case (see pp. 143–4). But in *350.44 and *350.51, Russell and Whitehead showed that they were satisfied in all open and connected families.

The two last conditions set out strong constraints over the relation between the addition and multiplication between ratios and the relative product of κ-vectors. Demand (3) concerns the 'change of unit' and states that, R, S and T being any three κ-vectors, $\mathbf{M}_T(R) = \mathbf{M}_S(R) \times \mathbf{M}_T(S)$. That is, the measure of T according to the unit R should be the product of the measure of R according to the unit S with the measure of S according to the unit T. Finally, condition (4) is about the addition of measures. It requires that $\mathbf{M}_S(T \mid R) = \mathbf{M}_S(T) + \mathbf{M}_S(R)$: the measure of the relative product of two vectors is the addition of their measures.[32] In *351.31, Russell and Whitehead showed that condition (3) is satisfied in any open connected submultipliable family κ, and they proved in *351.43, that if, in addition, κ contains all the powers of its vectors, then (4) is also fulfilled.

When pure ratios are restricted to measurable families, they become applied ratios. Applied ratios have the same arithmetical properties of the pure rational numbers, but, in addition, they can be regarded as measures of quantities – in the sense that they satisfy the four conditions given above. In Russell (1900b),[33] as in Hale's Fregean construction, the rational numbers were directly defined measures of quantities. The two kinds of properties distinguished in PM (the pure arithmetical properties and the features connected to measurement of quantities) were not disentangled: the arithmetical properties were inherited from the formal structure of the q-domains. The real strength of Russell and Whitehead's approach lies in the fact that what was regarded in Russell (1900b) or in Hale (2000) as a starting point (the structure of distance or of full q-domain) is now considered as the final stage of a long development. This more general perspective gives us the means to distinguish the 'pure' from the 'applied' ratios, while preserving the connection between numbers and their use in measurement (by satisfying *Applic*). Before developing this point, let me say a word about the PM theory of measurement by real numbers, expounded in the rest of section C.

Note first that the definition of measurable family does not say anything about real numbers. Condition (2) above makes it clear that, to be measurable, a family must be such that a unique vector should correspond to any positive rational number – real numbers are here not even mentioned. Now, according to *Applic*, the real numbers must be linked to their application in measurement. How then did Russell and Whitehead connect real numbers and magnitudes in section C? The section *356 devoted to the 'measurement by real numbers' lies at the conclusion of a difficult development (filling sections *352 to *355) aimed at generalizing the concept of the Möbius net. Of course, all the geometrical surrounding is, at this stage of PM, wanting. However, all this happens as if the authors had tried to define the minimal conditions which would allow them to give a sense to the problem of the introduction of coordinates.[34] To understand their theory of measurement by a real number, one must first grasp the main lines of this development.

After having defined the concept of measurable family, Russell and Whitehead introduced the concepts of 'rational family' and of 'rational net' (*354 and *355). A 'rational family' is one in which every vector is a rational multiple of some one unit S. A 'rational net' is a more complicated notion. Let's first select, among a given family κ, all the rational multiples of a certain vector S. The subfamily κ' so constructed will be a rational family. Now, let's restrict the field of the κ-vectors to the α-points accessible from a certain point a through the κ'-vectors. The set so obtained will be a subset α' of the original domain α. The family κ' defined on α' is a rational net – more precisely, a $(\kappa, S\ a)$-rational net since it depends on κ, S and a. A rational net is then the result of the application to any vector-family of two successive operations of selection: the picking of the rational multiples of a certain κ-vector S, and the restriction of α to the points accessible from a given α-point a through the multiples of S.

In *355, Russell and Whitehead studied the relations between κ and the rational nets one can extract from κ. For instance, they showed that if κ is open, connected and close under the relative product, and if a is a connected point, then every (κ, a)-rational net is transitive, connected, open and close under the relative product, as well as transitive and asymmetric.[35] All this development, to which I cannot do justice here, clearly deals with the issue related to the introduction of coordinates on the projective line. As we saw in Chapter 1, the key problem the followers of Von Staudt were confronted with was to characterize the relation between the Möbius net and the line on which it was defined. In PM *355, Russell and Whitehead came back to this issue, which they now addressed in the very general terms of their theory of the measurable vector-family. In this respect, their analysis should be compared to the one developed by Veblen and Young (1910; 1914), who attempted to characterize the various species of projective lines (real, complex, finite projective lines) by setting some constraints on the relations between the line and the Möbius nets it supported. Here as well, the approach is very general. But it seems that Russell and Whitehead went one step further in completely severing this question from any reference to a geometrical setting.[36]

We are now in a position to tackle the issue of measurement by real numbers. The subject is indeed only a particular case of the general issue concerning the relations between a family κ and its rational nets. If κ is a family which is ordered, then it is possible to give a meaning to the notion of a limit of a sequence of κ-vectors; and, as an order can always be introduced on a rational net (via the rational coordinates), one can also give a meaning to the notion of a limit of a sequence of vectors belonging to a κ-rational net. This enables us to raise the following question. What are the relationships between the κ-vectors and the limit of sequences of vectors belonging to the κ-rational nets? This is the problem Russell and Whitehead addressed in PM *356. And it is just an abstract reformulation of the old Von Staudtian puzzle. We saw in Chapter 1 that Von Staudt did not succeed in proving that, on the real projective lines, the points with real coordinates could be introduced as limits of the sequence of 'rational' points (of points belonging to a Möbius net). The question this failure raised was: how can we characterize the relationship between the limits of the sequences of 'rational' points and the points of the real projective line? One recognizes here a particular form of the problem raised in *356.

Russell and Whitehead's answer was unsurprising: 'if a given set of vectors, all of which are rational multiples of a given vector R, have a limit with respect to U_κ [the order relation on κ], and if their measures determine a segment of $\mathbb{Q}^+$, then we take the real number represented by this segment as the measure of the limit of the given set of vectors' (1913, p. 442). In other words, and by oversimplifying Russell and Whitehead's development, real coordinates are introduced as limits of a sequence of 'rational' points. The two philosophers proved that, when κ is serial, submultipliable and semi-Dedekindian,[37] the real measures satisfy the four conditions enumerated at the beginning of section C. As soon as their field is restricted to suitable families, real numbers then become applied numbers (measures of vectors).

There is still a difference, however, between measurement by a real number and measurement by a rational. In the former case, a limiting process is essentially involved: a unit being fixed, a real measures 'the limit of [a] set of vectors', while a rational measures one vector only. The 'nested' feature of PM real numbers, noted at the end of section 5.2, is simply a consequence of this difference. Indeed, when no account is taken of the special sense in which real numbers measure quantity, one is led to maintain that two magnitudes can have many different real numbers (in the sense that they belong to many nested sets of couples representing real numbers); but as soon as a real measurement is defined, as it should be, as a limiting process, then, if S is measured by a real number X, there is no other real number Y which measures it. Of course, one could object to this approach that it is needlessly complicated: in Frege-Hale's perspective, for instance, there is no difference in the ways rationals and reals measure the quantities belonging to a complete q-domain. Why then distinguish between rational and real measurements? In PM, this distinction directly proceeds from the connection drawn between the measurement by real numbers and the introduction of real coordinates. In such a context, the rational case enjoyed a very special status: the rational net was defined by the iteration of a fundamental operation (the product of a relation by itself in PM, the quadrilateral construction in Von Staudt), and it was an indispensable instrument for studying the properties of the structure which supports it (the vector-family κ, the projective line). This is why I have put here so much emphasis upon the geometrical context of Russell and Whitehead's reasoning – to be intelligible, the theory of measurement by real numbers should be seen as a chapter of a general analysis of the relationship between a vector-family and its rational nets.[38]

5.4 Hale's Fregean axiomatic definition versus Russell and Whitehead's relational theory

Let me now summarize the general structure of Russell and Whitehead's theory. In section A, a pure theory of rational number is developed, independently of any consideration relative to quantity. In section B, a general theory of magnitude is presented that is not connected with measurement. The only connection between the two first sections is that numbers are defined as relations of relations, and that quantities are introduced as relations. It is only in section C that the two first theories are articulated. Thus, the organization of PM VI can be represented as in Figure 5.1.

As I stated at the outset, Russell and Whitehead have the same goal as Frege: they explicitly seek a definition of numbers which would account for the use of them in measurement of quantities, without implying the sort of physicalism assumed by Burali-Forti. Still, when compared with Frege's account, Russell and Whitehead's conception seems at first needlessly complicated. Recall that in Hale, the definitions of q-domain and of ratio lead very easily to a satisfactory characterization of the rational (and real) numbers. Thus, instead of the intricate schema of Figure 5.1, one ends up with a diagram of the kind shown in Figure 5.2.

A mere glance at the two figures reveals the first to be much more involuted than the second. In particular, in Figure 5.1, two appendages occur, which seem to have no use. Why develop a theory of pure numbers (represented

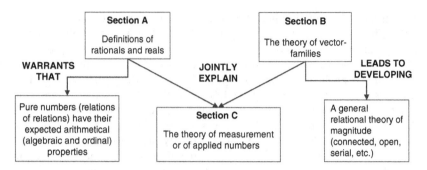

Figure 5.1   The structure of PM VI

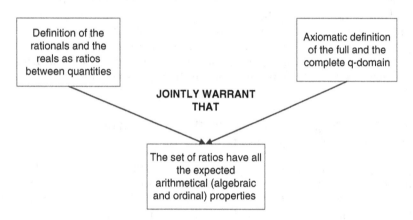

Figure 5.2   Frege's doctrine

by the appendage at the left) and a general relational doctrine of quantity (represented by the appendage at the right)? Why complicate that which can be done more simply?

I would like to show that these apparently needless complications make PM's theory more natural and straightforward than Hale's Fregean theory. Let me first focus on the left side of Figure 5.1, that is, on the definition of pure numbers. In Hale's and Frege's theory, in order to guarantee that the rationals (the reals) have the expected properties, one must show that there is at least one full (complete) q-domain. This proof is not straightforward. It has been shown that Hale has to construct a full (complete) domain, with the employment of certain abstraction principles, from the domain of natural numbers. Now, in PM VI, no extra condition on the structure of the field to which the ratios are applied is needed to guarantee that the ratios have their standard arithmetical properties. The mere fact of characterizing a ratio as a relation between relations suffices to endow it with all the required mathematical structure. Relations, unlike quantities, are logical entities in PM. One can thus rely on logic to show that there are sufficiently many relations for endowing numbers with their usual arithmetical features.

This appendage concerning the theory of pure numbers, far from being an idle device, exhibits the logical nature of Russell and Whitehead's theory of rational numbers. And it exhibits this more straightforwardly than Frege's theory. Since Hale has to resort to Dedekind's cut construction to immunize his theory against physicalism, his account is open to the following obvious objection: if cuts (or other equivalent devices) are required after all, then why not just identify reals with cuts? In Russell and Whitehead's theory, all these twists and turns are eliminated immediately by the assumption that numbers connect relations. The introduction of the concept of relation, and of the logical machinery which accompanies it, enables them to develop a theory of pure numbers and then to dispense with the artificial construction of a particular model of the relevant q-domain.

Let me now focus on the right side of Figure 5.1, that is, on the general theory of vector-families. Frege and Hale claim that their theory of numbers explains the application of numbers in measurement. But at the same time, they develop a very restricted account of application, based on the notion of a full (complete) q-domain. Now, in the actual practices of scientific measurement, numbers are used to measure many different kinds of quantities, some of which have nothing to do with extensive magnitudes. Starting from a measure-theoretic perspective, Batitsky (2002) gives many examples of scales which have nothing in common with Hale's q-domain. Consider the ordinal scales, as the Mohs scale used to measure the hardness of stones. It is clear that Hale's definition of numbers in terms of ratios does not provide any clues as to how numbers are used in ordinal scales. This suggests that in order to follow Hale in his claim that his definition, unlike that of Dedekind, can explain how numbers are used in measurement, one has to grant a certain kind of privilege to the measurement in full (complete) q-domains. This is precisely what Batitsky objects to: from a measure-theoretic point of view, extensive scales are just one among many possible sorts of scale and there is nothing which justifies the pre-eminence which Hale intends to give them. An additional argument is then needed to establish that a full (complete) q-domain is the most fundamental sort of quantity – a consideration which complicates Frege's and Hale's construction.[39]

Importantly, Russell and Whitehead avoid this pitfall, since their theory of magnitude in section B is general and completely independent of any definition of number. As we have seen, magnitude or vector-family appears in PM to be a very weak formal structure, which can be enriched in many different ways, and which can accommodate many of the most bizarre scales of modern measurement theory. The vector-family contains an 'additive' operation and, in this respect, it is true that a purely ordinal scale (like the Mohs scale) cannot, strictly speaking, be interpreted as a vector family. Recall, however, that the composition operation is not necessarily closed, and that the theory of the serial family does not rely on the existence of this operation. Thus, even an ordinal scale could, with some emendations, be inserted into the frame of Russell and Whitehead's theory.[40] Batitsky's criticism concerning the narrowness of Frege's and Hale's conception of quantity does not arise in PM. The general theory developed in section B prevents the objection from the very beginning.

The two apparently needless 'appendages' of Figure 5.1 are not useless, then. While preserving the kernel of Frege's project (combining the logicist requirements with the need to account for applications), they allow Russell and Whitehead to avoid the two traps into which Frege and Hale fall. To push the analysis of the contrast further, it is worth considering what the genuine philosophical source of the difference between Hale's Fregean theory and Russell and Whitehead's theory might be. In particular, it is worth considering what prevents the former approach from developing the same sort of solution provided by the latter view.

In Hale and Frege, *Applic* is conceived in structural terms. To secure density among the ratios, one needs to apply them to a densely ordered q-domain. Conversely, only q-domains whose ratios are susceptible to generating numerical structures akin to the rational or real fields are considered. Shapiro has criticized Hale's recourse to *Applic* from the structuralist perspective.[41] However, the way Hale implements *Applic* has a structuralist flavour. The basic idea underlying his, Frege's and even Burali-Forti's construction is to view numerical and quantitative structures as isomorphic. If numbers can be applied to quantities without losing their arithmetical properties, it is because of the formal shapes of the q-domains. The problem is that when one connects so tightly the structure of quantities to the structure of numbers, one is likely to lose on both sides. One is compelled to restrict drastically one's theory of magnitude, so as to recover the arithmetical properties of numbers;[42] and one is led to loosen one's grip on the insight that the arithmetical properties of numbers do not depend at all on the existence of a relevant sort of quantitative domain.

The authors of PM endorse a completely different reading of *Applic*. Russell and Whitehead's main insight is that the connection between numbers and quantities should be made at the ontological or typological level, not at the structural level. What makes numbers specially applicable to quantities is not the fact that numbers and quantities share the same abstract structure, but the fact that their logical types are adjusted to each other: quantities are relations and numbers are relations between relations. Locating the connection between numbers and quantities at this level gives extra space to the logical and mathematical analysis. Since the structural shapes of the q-domains are not constrained by the arithmetical features one seeks to recover, it is possible to elaborate a general theory of quantity, free from any connection with numbers. And likewise, since one defines ratios as relations between relations (instead of relations between abstract quantities), one is able to get all the required arithmetical results directly from logic, without putting any constraint on the theory of quantity.[43]

This does not mean, however, that PM's doctrine does not give rise to any difficulties of its own. In Hale's (as in Frege's and Burali-Forti's) theory, there is an effort to remain faithful to the surface grammar of quantitative statements. Thus, Hale does not say anything about the inner nature of quantity; he just attempts to characterize the relations between elements of a q-domain. This is in line with his structuralist strategy: quantities are for Hale bare 'places in the structure', placeholders which can be filled by anything which possesses the relevant properties.[44] Quantities can be relational, but they do not have to be so. In Russell and Whitehead's view, by contrast, quantities are not 'bare places'; they are ontologically more determined. In PM, quantities are relations. This raises a difficulty, since 'we commonly think of ratios as applying to magnitudes other than relations' (1913, p. 260). Mass, for instance, seems to be the property of bodies, if anything – and not relations. The fact that Russell and Whitehead went beyond a mere structural characterization of quantity compelled them to distort the surface grammar of quantitative statements (1913, pp. 260–1):

> In applying our theory to (say) the ratio of two masses, we note that the idea of quantity (say, of mass) in any usage depends upon a comparison of different quantities. The 'vector quantity' R, which relates a quantity m_1 with a quantity m_2, is the relation arising from the existence of some definite physical process of addition by which a body of mass m_1 will be transformed into another body of mass m_2. Thus σ such steps, symbolized by R^σ, represents the addition of the mass $\sigma(m_2 - m_1)$ ... Thus to say that an entity possesses μ units of quantity means that, taking U to represent the unit vector quantity, U^μ relates the zero of quantity – whatever that may mean in reference to that kind of quantity – with the quantity possessed by that entity. It can be claimed for this method of symbolizing the ideas of quantity (α) that it is always a possible method of procedure whatever view be taken of it as a representation of first principles, and (β) that it directly represents the principle 'No quantity of any kind without a comparison of different quantities of that kind.'

Russell and Whitehead explained here how to reparse quantitative propositions about mass so as to make them appear relational.[45] In Hale's perspective, such a reformulation is not needed at all, and, in this respect, one could consider that the structuralist strategy has an advantage over the approach developed in PM. But Russell and Whitehead's move is here part of a larger programme, launched in *On Denoting* and perhaps even before: if it can sometimes be a guide for logical analysis, grammar should never be followed blindly. That the surface form of quantitative statements is thus not relational does not mean that their logical form is not genuinely relational. And the PM relational theory of quantity is, in this respect, just a (superb) illustration of the idea 'that the apparent logical form of the proposition needs not be its real form' (Wittgenstein 1921, 4.0031). This connection between the PM way of implementing *Applic* and the distinction between apparent and real forms of proposition is absolutely crucial. I will come back to this topic in the next chapter.

What emerges from the comparison between Hale (2000) and PM is that the application principle is interpreted in a completely different way in the two works. Hale's strategy is structural (*Applic* is satisfied by an adjustment of the numerical and the quantitative structure), whereas Russell and Whitehead's approach is ontological (*Applic* is fulfilled by an adjustment of the types of number, characterized as a relation of relations, and the type of quantity, characterized as a relation). Frege, like Russell and Whitehead, recognized the legitimacy of two distinct claims: the logicist one, according to which the arithmetical properties of numbers do not depend on a physical fact; and *Applic*, according to which numbers should be connected with their use in measurement. My point is that PM strategy offers a better framework for articulating these two divergent demands. Hale tries to unify the structure of numbers with that of quantities, but, in so doing, he loses sight of the differences between the two concepts. Russell and Whitehead were more sensitive to these differences and, instead of seeking to unify numbers and magnitudes, they looked for a framework in which the two structures could be integrated without losing their formal specificities.

5.5   Conclusion

My first aim in this chapter was to show that the letter Whitehead sent to Russell on 14 September 1909 faithfully described the main lines of PM VI. Russell and Whitehead's theories of rational and real numbers were not a mere logical rewriting of Dedekind's construction. On the contrary, as had Burali-Forti and Frege, the two logicists wanted to satisfy *Applic*; and they drew a tight connection between numbers and measurement. At the same time, Russell and Whitehead, like Dedekind and Frege, but against Burali-Forti this time, shared the view that the arithmetical properties of numbers do not depend on some empirical facts. Thus, the authors of PM attempted, with Frege, to construct an intermediate path between Dedekind's arithmetical definition and Burali-Forti's physicalist characterization. But the way they achieved this task was very different from Frege. Instead of grounding their definition of numbers on a formal definition of quantitative domains, Russell and Whitehead adjusted the type of rational and real numbers (defined as relations of relations) to the type of quantities (defined as relations). I have here tried to give a survey of the main stages of PM VI's intricate doctrine. Even if what precedes does not do full justice to the wealth of Russell and Whitehead's development, I hope I have said enough to convince the reader that the way PM VI implemented *Applic* represents an alternative which, at least, stands comparison with Hale's recent Fregean construction.

A question remains open, however. One thing is to show that the PM way of implementing *Applic* is a good one; another is to justify the use of *Applic* itself. I have fulfilled the first task, but, until now, I have said nothing about the second one. Why did Russell and Whitehead claim that a good definition of rational and real numbers should account for their application in measurement? In 1903, Russell endorsed the arithmetical definition of rational and real numbers without letting scruples concerning application get in his way. PoM III expounds an elaborated doctrine of quantity, but Russell made it clear that this theory was only a 'concession to tradition' that did not play any role in the definition of numbers. Why did he change his mind in PM? In the next chapter, I will try to answer this question. My suggestion will be that PM's adhesion to *Applic* should be linked to the idea that grammar does not guide the analysis. Thus, I will argue that the rejection of the structuralist outlook has not only some consequences on the way *Applic* is implemented, but that it is at the source of Russell and Whitehead's endorsement of *Applic*. The grammar being not a reliable guide, *Applic* played the role of a guiding principle governing analysis.

6
Application Constraint in *Principia Mathematica*

In the previous chapter, we saw that Russell and Whitehead, like Frege, endorsed *Applic*, i.e. the principle according to which a good definition of a mathematical concept (rational and real numbers in the case in point) should account for its main applications. I emphasized that the way they implemented the constraint was different from Frege's, but I have not yet explained why they endorsed *Applic*.

Applic seems to be composed of two distinct theses:

1. The recognition that there is a distinction between the mathematical content of a concept and its extra-mathematical uses.
2. The thesis that a good definition must relate the concept to its main extra-mathematical uses.

In Chapter 3, we saw that Russell's anti-psychologism is based on the claim that one should not introduce, in the logical content of a mathematical concept, any extraneous elements. Point (1) describes the application as external to the mathematical content of the concept considered, and point (2) requires that one should nevertheless regard it as an integral part of the defined concept. What *Applic* asks for seems to be something that the attack against psychologism prohibits. Of course, the clash between Russell's anti-psychologism and *Applic* is not as straightforward as I suggest here. Considerations about application have prima facie nothing to do with the psychology of mathematicians. But the idea that certain non-mathematical elements should be included in the mathematical content inevitably triggers the anti-psychologist reaction. We have also seen that Russell very much extended the scope of his psychology – all human centred considerations come within its province. Now, some tenets of the story (the emphasis on the fact that *we* use numbers in measurement) told by the upholders of *Applic* seem rather to be a matter for a theory of human activity than a matter for an account of the mathematical content. If *Applic* leads us to blur the distinction between logic and psychology, why then did Russell and Whitehead adopt it?

The aim of this chapter is to show that the PM use of *Applic* does not conform to the two-stage construction I have just expounded. Russell and Whitehead rejected point (1) and then point (2). Indeed, for the logicists, the boundaries of the mathematical concepts whose applications are considered were not fixed in a univocal way in the existing mathematics. Far from being regarded as a given datum, the division of the mathematical field into well-defined, distinct branches is a task that the logical analysis is supposed to perform. And *Applic* is instrumental in the fulfilment of this work: to look at different contexts of use is a means of probing different ways of carving out the mathematical content and to make definite the borders of the different disciplines. There is thus a feedback effect in the PM use of *Applic*: the examination of the applications of a particular concept gives information on its content. Russell and Whitehead did not see *Applic* as a principle allowing them to relate two already separated fields of knowledge, but as a device enabling them to locate the joints at which mathematics must be carved. In what follows, I will attempt to describe the specificities of the PM use of *Applic* and spell out its many consequences.

In the first section, I examine a possible justification for *Applic* that Whitehead might have adopted, but that Russell could not accept: the metaphysical claim that, in mathematics, there is always a fit between the most abstract and general concepts and the most concrete applications. In section 6.2, I focus on the articulation between PM VI (the theory of real numbers) and PM V (real analysis, i.e. the theory of limit and convergence). In 1913, real analysis and the theory of real numbers were for the first time distinguished, and I will show that, when put in its historical context, this move turns out to have great significance. In section 6.3, I come back to *Applic*. This is instrumental in the distinction between real analysis and real numbers, and this use shows that, in PM, *Applic* is not a demand to relate an already known branch of mathematics to a different field, but a means of investigating the outlines of the theoretical body whose applications are considered. In section 6.4, I turn to the recent discussion on *Applic*, and contrast Russell and Whitehead's interpretation with the one developed by Shapiro, Hale and Wright. All these philosophers make a sharp contrast between the epistemological task of identifying the mathematical structure (for instance, the structure of $\mathbb{R}$) one wants to define and the metaphysical task of characterizing the ontological nature of the objects which populate it (in the case in point, the nature of real numbers). I claim that this distinction is completely at odds with the PM understanding of *Applic*: for Russell and Whitehead, *Applic* is used to meet the epistemological demand and not just to determine the metaphysical nature of the already characterized mathematical structures. In section 6.5, I connect this discussion with the developments expounded in Chapter 3. In particular, I hold that, in

PM, *Applic* is not a mechanical device allowing us to uncover the real logical form of a given concept – it is more like a guiding rule of thumb, which should be adjusted to the particular features of the conceptual landscape one is surveying.

6.1 Whitehead's conception of applied mathematics

PM VI was written by Whitehead. It was thus Whitehead who offered Russell a departure from the orthodox arithmetical view. One could suppose that Whitehead had special reasons for doing this. I do not know of any passage where the philosopher explains why he endorsed *Applic*. But in both his early and late works, Whitehead developed a view about the harmony between abstract thought and concrete applications which may have led him to *Applic*.

As Desmet (2010, pp. 97–8, 165–6) has explained, Whitehead, like many of his contemporaries, had been deeply impressed by Maxwell's theory of electromagnetism. In particular, he seemed to have been struck by the fact that Maxwell's unification of electricity and magnetism was rooted in Hamilton's quaternions. Let me quote a passage from *An Introduction to Mathematics*, where Whitehead spoke about complex and hypercomplex number systems (1911, p. 100):

> One of the most fascinating characteristics of mathematics is the surprising way in which the ideas and results of different parts of the subject dovetail into each other. During the discussions of [complex number systems] we have been guided merely by the most abstract of pure mathematical considerations; and yet at the end of them we have been led back to the most fundamental of all the laws of nature, laws which have to be in the mind of every engineer as he designs an engine and of every naval architect as he calculates the stability of a ship. It is no paradox to say that in our most theoretical moods we may be nearest to our most practical applications.

The generalization of number (the process by which we go from natural numbers to complex and hypercomplex numbers), even if it is driven by purely formal reasons (like the principle of 'permanence of form'), is not cut from concrete applications. For Whitehead, the work of Maxwell showed that 'in our most theoretical moods we may be nearest to our most practical applications'.

This idea is not abandoned after 1913.[1] Hence, in *Science and the Modern World*, Whitehead explained (1925, pp. 31–2) that:[2]

> The birth of modern physics depended upon the application of the abstract idea of periodicity to a variety of concrete instances. But this would have been impossible, unless mathematicians had already worked out in the

abstract the various abstract ideas which cluster round the notions of periodicity. The science of trigonometry arose from that of the relations of the angles of a right-angled triangle, to the ratios between the sides and the hypotenuse of the triangle. Then, under the influence of the newly discovered mathematical science of the analysis of functions, it broadened out to the study of the simple abstract periodic functions which these ratios exemplify. Thus trigonometry became completely abstract; and in becoming abstract, it became useful. It illuminated the underlying analogy between sets of utterly diverse physical phenomena; and at the same time it supplied the weapons by which any such set could have its various features analysed and related to each other ... Nothing is more impressive than the fact that as mathematics withdrew increasingly into the upper regions of ever greater extremes of abstract thought, it returned back to earth with a corresponding growth of importance for the analysis of concrete fact.

Note that Whitehead did not attempt to explain this 'fit' between abstract thought and concrete phenomena.[3] For him, this 'impressive' adjustment is just a fact, which we have to register and which should be brought forward in any philosophical reflection about mathematics. In this respect, Whitehead's position seems to be close to the notorious one expounded by Wigner (1960). For the two thinkers, the effectiveness of mathematics in the natural sciences is unreasonable.[4]

The idea that abstract generalization in mathematics 'miraculously' coincides with the most concrete applications (and that one should never lose sight of this when dealing with mathematics) is a view that Whitehead entertained throughout his philosophical career. Partly because it is related to the material dealt with in Chapters 1 and 2, I would like to add more evidence of the importance of this insight in Whitehead's early thought. One of the leitmotivs of *Universal Algebra* (1898, p. 32) is that 'a treatise on... Algebra is also to some extent a treatise on certain generalized ideas of space'. For Whitehead then, generalization in algebra (the most abstract part of mathematics) went hand in hand with the development of geometry (the most concrete part of mathematics). Let me briefly explain the idea.[5]

By 'universal algebra', Whitehead essentially meant the various vector calculi expounded in Grassmann's second edition of the *Ausdehnungslehre* (1862).[6] Now, in this work, Grassmann only considered the application of his 'calculus of forms' to Euclidean geometry (p. 123 ff.). This restricted standpoint led him to put certain ad hoc constraints on the interpretation of his algebraic forms. Thus, the 'intensity' (the scalar) associated with a linear combination c of two given 'vectors' e_1 and e_2 ($c = a_1e_1 + a_2e_2$, where a_1 and a_2 are real numbers) was defined to be the sum of a_1 and a_2 (ibid., pp. 125–6). To understand this choice, think of e_1 and e_2 as two points on a given Euclidean line. Now, to each couple (a_1, a_2) of real numbers, one can associate the point c such that the ratio of the (signed) distance from c to e_1 over the (signed) distance from c to e_2 is a_2/a_1 (when a_2 and a_1 are both positive, the point c can be viewed as the barycentre of the system containing the point e_1 of mass a_1 and the point e_2 of mass a_2). For instance, when a_1 is equal to a_2, c is the middle point between e_1 and e_2. One can show that, when $a_1 + a_2 \neq 0$, there is a unique point on the Euclidean line (e_1, e_2) satisfying the demand. But if $a_1 + a_2$ is equal to 0, there is no Euclidean point fulfilling the said condition – there is no point on the line (e_1, e_2), outside the segment $[e_1, e_2]$, equidistant to both e_1 and e_2. In this case, Grassmann says that c is not a point, but a displacement[7] lying at infinity on the Euclidean line.[8]

In other words, to construct a geometrical interpretation of his algebra, Grassmann was forced to restrict the generality of the calculus and to add an ad hoc postulate which stipulates that the value of the intensity of a linear combination c of two vectors e_1 and e_2 is the sum of the intensities of e_1 and e_2. As Whitehead explained (1898, p. 168):

> Grassmann does not consider the general question of the comparison of intensities. In the *Ausdehnungslehre von 1844*, 2nd Part, Chapter 1, §§ 94–100, he assumes in effect a linear defining equation without considering any other possibilities. In the *Ausdehnungslehre von 1862* no general discussion of the subject is given; but in Chapter V, 'Applications to Geometry', a linear defining equation for points is in effect assumed . . . assumptions which are obvious and necessary in Euclidean Geometry.

In the last chapter of the third book (1898), Whitehead showed how an appropriate generalization of the geometrical setting gives us the means to connect directly Grassmann's algebra with the various (classical) non-Euclidean spaces. Indeed, instead of restricting right away the intensity function (linking the intensity of c to that of e_1 and e_2) to a linear form, Whitehead started from a very general 'intensity law' and shows that, when this law takes the form of a quadratic equation,[9] then the various loci of zero intensity correspond to Cayley's Absolutes. In book VI (1898), the connection between Cayley-Klein's projective definition of metric and the Grassmannian algebra is fully worked out:[10] the three classical metrics (elliptic, hyperbolic and parabolic) are characterized by a particularization of the quadratic 'intensity laws'.

Whitehead (1898) remained faithful to Grassmann's project of developing a theory of pure extension;[11] but to achieve this task, Whitehead realized that he had to break with Grassmann's exclusive focus on Euclidean space. Indeed, according to him, the reason why Grassmann needed to restrict the generality of his calculus was not that his algebra was too abstract. Restrictions were needed because geometry was, at the time, not developed enough to provide us with the right interpretative framework of Grassmann's theory. Whitehead's reading of the *Ausdehnungslehre* thus epitomizes the idea 'that branches of mathematics, developed under the pure imaginative impulse ... finally receive their important application' (1928, p. 6). Of course, geometry was not considered in 1898 to be an empirical science and, properly speaking, the connection between universal algebra and projective space cannot be taken as an illustration of the fit between abstract thought and concrete applications within mathematics. Yet this analysis constitutes a neat illustration of Whitehead's idea that the different parts of mathematics 'dovetail into each other'.

Let me now come back to our initial question about the justification of *Applic* in PM. Since *Applic* is nothing other than the demand to reflect in the definition the fact that the abstract mathematical concepts are tied up to concrete applications, one can well understand why Whitehead could have found it attractive. To define real and rational numbers as relations of relations, and to adopt correlatively a relational theory of magnitudes, enabled him to illustrate, once again, his general standpoint about the deep unity of our thought. From this perspective, Dedekind's theory of reals, which severs numbers from measurement, if formally correct, would be philosophically unsatisfying since it would not display the intimate connection between abstract ideas and concrete applications. As I said, I do not know any text where Whitehead justifies the PM definition of rational and real numbers in these terms. Such reasoning is however in line with what the philosopher constantly said in all his works.

For at least two reasons Whitehead could not hope, however, to convince Russell with this argument. First, PM officially aims at being philosophically neutral. In the Preface, the two authors wrote (1910, p. v):

> We have ... avoided both controversy and general philosophy ... The justification for this is that the chief reason in favour of any theory on the principles of mathematics must always be inductive, i.e. must lie in the fact that the theory in question enables us to deduce ordinary mathematics.

If anything is, the claim that there is always a fit between abstract thought and concrete applications is a metaphysically loaded thesis. As we have seen, Whitehead did not even provide any explanation for this 'fact'. To justify *Applic* by saying that the principle illustrates the deep unity, in our thought, between formal developments and applications, was then not an option. Moreover, it seems that Russell did not share Whitehead's metaphysical belief.[12] Philosophical neutrality was thus not just a methodological position – it was required to avoid any conflict of opinion between the two friends. And there is another reason which precludes Whitehead's argument from playing any role in PM. PM III (concerning cardinal arithmetic) and PM IV (concerning relation arithmetic) are variations on Cantor's transfinite arithmetic, a part of mathematics which was, at the time, criticized for being an empty formal game without any application to 'real' mathematics (to say nothing of its application to the physical sciences).[13] To claim that all the main mathematical concepts and theories are intimately connected to concrete applications would be shooting oneself in the foot: some parts of PM are full of abstract developments, which have absolutely no applications in the empirical sciences (and even in ordinary mathematics).

The belief that there is a fit between abstract thought and practical applications has certainly played a role in Whitehead's critical reaction to Dedekind's definition of real numbers. But it cannot be considered as a justification of *Applic* in PM. Russell would never have bought such a metaphysical argument – which moreover would have been at odds with his reception of Cantor's works. We are thus led back to our initial puzzle: what are the reasons which led not only Whitehead, but also Russell, to take applications into account in PM VI?

6.2   Real analysis and real numbers

To make some progress, I will attack the issue from a completely new angle and study in more detail the way PM VI is articulated with the rest of the book. The published PM is divided into six parts: the first one is devoted to mathematic logic (ramified type theory, theory of description, theory of classes); the second and third are dedicated to cardinal arithmetic (equivalence relations, addition and multiplication between cardinals, finite and infinite sets); part four is a generalization of Cantor's ordinal arithmetic (called 'relation arithmetic'); part five is concerned with series and real analysis (definitions of limit, convergences, continuity); and part six is concerned with numbers and quantities. This organization sustains PoM's main division. In PoM, the first part is devoted to logic; a distinction is made between cardinal and ordinal arithmetic in Parts II, IV and V; and a detailed study of the foundations of real analysis is developed in Part V. If one takes into account the fact that the fourth (never published) volume of PM was supposed to deal with geometry, the fit between the two works is almost perfect.[14] Almost – since, as I have already had occasion to note, the theory of quantity does not play the same role in PoM and PM (1913). The doctrine of magnitude was a 'concession to tradition' in PoM, whereas the theory of the vector-family is an essential part of the doctrine of rational and real numbers in PM.

To account for this discrepancy, I will study in more detail the PoM and PM accounts of the foundations of real analysis. At first, it seems that the two conceptions do not differ. Thus, in PoM, chapter 33, Russell, after having defined the notion of a segment of rationals (p. 271), identified a real number with a segment ('a segment of rationals *is* a real number' (p. 272)). In PM, the notion of a segment is still considered as a key concept: in *275, after having characterized the series of order type η (the 'rational' series)[15] and the series of order type θ (the 'continuous' series),[16] Russell and Whitehead, following PoM closely, proved that the series of segments of a rational series is a continuous series (*275.21). This proposition is merely a reformulation of Dedekind's key result. How can we combine, then, this use of the Dedekindian toolkit in PM V with the 'anti-arithmetical' perspective on rational and real numbers developed in PM VI? Would it not have been better to espouse the PoM point of view and to expel the theory of quantity from the foundation of real analysis?

But the resemblance between the PoM and PM accounts of real analysis is in fact not perfect. The essential point is that, in PM, a distinction is made between real analysis and real numbers, while in PoM the two notions are confused. Let me explain this point by focusing first on PoM.

After having defined real numbers in terms of segments of rationals in chapter 33, Russell discussed the other options. He first dealt with the definition of reals in terms of limits (chapter 34) and then with the characterization of reals in terms of distances (chapter 35). In the course of his argumentation, he is led to maintain that 'continuity [is] a purely ordinal notion' (PoM, p. 296). Thus the conclusive chapter 36 presents a purely ordinal account of continuity: Russell first defined (PoM, p. 276) the order type η and the order type θ in a purely ordinal way, and then constructed a continuous series from a series of type η in the same manner, as is explained in PM *275.21.[17] The reasoning expounded in chapter 36 is then a general description of the construction expounded in chapter 33: the derivation of real numbers from rational ones appears now as just a particular case of a purely ordinal process. This raises a question however: what is the status of the real numbers in PoM? At the end of the development, the real numbers are just an example of the more general concept of a continuous series. As Russell himself noted, 'it must not be supposed that continuity as above defined can only be exemplified ... by the devious course from integers to rationals, and thence to real numbers' (PoM, p. 296). Now, according to Russell, this is the ordinal notion of continuity which plays a foundational role in real analysis, i.e. which is intimately connected to the other fundamental concepts like limit and convergence. Nothing in the PoM account of real analysis depends on the particular example of continuous series that the real numbers are. But if so, why did Russell introduce real numbers in chapter 33?

A parallel with arithmetic could be helpful at this stage. In PoM, Russell recognized that one can base finite arithmetic on the notion of progression. However, he refused to do so because grounding an account of the integers on Peano's axioms would lead to a severing of the whole numbers from their application in counting.[18] In the case of the whole numbers, Russell gave an argument explaining what makes the progression made up of finite cardinals so special. But, in PoM V, he supplied no similar reason to explain why one should study the particular continuous series that real numbers are.

In PM, the situation is much clearer. Rational and real numbers do not occur at all in PM V;[19] they are introduced only later, in PM VI, by reference to measurement. I have explained at length, in the last chapter, how Russell and Whitehead linked real numbers to relational magnitudes. But what I did not say there is that this new connection was the correlate of an ejection of numbers from the foundations of real analysis. This second move is not an unimportant detail. It is something that Whitehead and Russell explicitly prided themselves on having done. Let me quote the introduction of section C of PM V (1912, p. 715):[20]

> The purpose of this section is to express in a general form the definitions of convergence, the limits of functions, the continuity of functions, and kindred notions, and to give such elementary consequences of these definitions as may seem illustrative. In the definitions usually given in treatises on analysis, it is assumed that both the arguments and the values of the function are numbers of some kind, generally real numbers, and limits are taken with respect to the order of magnitude. There is, however, nothing essential in the definition to demand so narrow a hypothesis. What is essential is that the arguments should be given as belonging to a series, which need not be the same series as that to which the arguments belong. In what follows, therefore, we assume that all the possible arguments to our function, or at any rate all the arguments which we consider, belong to the field of a certain relation Q, which, in cases where our definitions are useful, will be a serial relation; we assume similarly that the values of our function, at least for arguments belonging to $C'Q$, belong to the field of a relation P, which, in all important cases, will be a serial relation.

In the remaining part of section C, Russell and Whitehead showed that, contrary to what is suggested in the 'definitions usually given in treatises on analysis', the main properties of convergence, limit and continuity can be proved without assuming that the arguments or values of the functions concerned are either numerical or numerically measurable – for a clear summary of PM V, section C, see Russell (1919, pp. 114–16). All that is required to prove the fundamental theorems of real analysis is that the arguments and the values of the functions considered belong to certain kinds of series. In PoM, Russell was aware of this point, for in chapter 36 real analysis is presented as a purely ordinal theory. But, as he was confused as to the relation between real number and continuous series, he did not oppose his ordinal conception of real analysis to the one that was usually developed 'in treatises on analysis'. For him, there was no real difference between real analysis and the theory of real numbers – and thus no real difference between providing a foundation to the former and providing an account of the latter.

To understand the importance of the change brought forward in PM, one should place it in its historical context. At the beginning of the twentieth century, spurred on by the works of Dini, Borel, Lebesgue and Baire, the theory of the real valued functions was a growing research field, which deeply interacted with Cantor's theory of ordinals. However, the mathematicians working in this area (especially the 'French analysts') usually regarded Cantor's theory as just a useful tool and not as an independent domain of research. Thus, the hierarchy of the Alephs (since it had no use in the theory of real valued function) was usually dismissed as an idle formal game – see Baire's reaction in footnote 13 above. This was yet not the sole way of receiving Cantor's work: during the same period, certain mathematicians developed the theory of ordinals for its own sake. Thus, in his 'Investigations into Order Types' (1906), Hausdorff explicitly set out a pure theory of order types, free from any consideration about its application to the doctrine of real valued function and to the theory of well-ordered series.[21] For Hausdorff, Cantor's theory was not a powerful instrument allowing an investigation of the topological structure of the subsets of $\mathbb{R}$, but a legitimate branch of mathematics on its own, which deserved to be explored as such. There was thus a complete opposition between the ways the French analysts and Hausdorff received Cantor's theory of ordinals.[22]

When inserted into this context, Russell and Whitehead's claim, according to which real analysis must be construed as a theory of continuous series, appears as an endorsement of Hausdorff's standpoint.[23] In PM, Cantor's theory of ordinals, suitably extended in a doctrine of order types,[24] supplies the framework for developing real analysis. In this perspective, real analysis does not appear any more as the study of the properties of certain numerical functions. Characterized as a doctrine of continuous series, real analysis has nothing to do with numbers. We saw that the French analysts considered real analysis as a chapter of the theory of real valued functions and Cantor's ordinal arithmetic as a useful tool to progress in this field. Russell and Whitehead turned this picture upside down: it is no longer the theory of real valued functions which occupies centre stage, but the doctrine of order; real valued functions, if useful from a didactic perspective, are just a particular case of a more general kind of function. What was considered as a tool (Cantor's theory of ordinals), whose end (the theory of real valued function, real analysis) was external to itself, became the framework within which real analysis should be embedded. In a certain sense, this reversal was already present ten years before, in chapter 36 of PoM. But, as we have seen, the insight was then parasitized by the fact that Russell gave a definition of rational and real numbers in chapter 33. All this happens as if Russell, in 1903, did not dare to endorse the full consequences of his ordinal theory of continuity.[25]

Let me come back to the main line of the argument. In PM V, real analysis is presented as a pure theory of order, not as a chapter of the theory of real valued function. Correlatively, in PM VI, a relational theory of real numbers is presented. The theory of real valued function is thus in PM divided into two independent doctrines: a purely ordinal theory of continuity, limit and convergence, on the one hand; and a relational theory of rational and real numbers, on the other. Why break up the unity of the theory of real valued function into two different subdisciplines? After all, as Russell and Whitehead acknowledged, to restrict real analysis to real valued function was at the time a standard view. Why then depart from the accepted conception? One here encounters a kind of question one has already met in this book. In many cases, there is more than one way of carving out the mathematical material, with no consensus in the mathematical community on the best way to do it. As we have seen, Russell in PoM was often confronted with the problem of choosing, among many alternatives, the articulation which corresponded to the real logical structure of the content. The present discussion supplies a new illustration of this fact. Russell hesitated in 1903 between two ways of presenting real analysis. In 1912, Russell and Whitehead clarified their thought: they defined real analysis as a chapter on the theories of series and real numbers as a chapter on a theory of measurement. But they could also have construed real analysis as a part of the theory of real valued functions. What were the reasons which led them to espouse the first option? How can we determine the best way to organize the mathematical field?

6.3   The architectonic use of *Applic*

It is at this stage that *Applic* plays a decisive role. Indeed, the separation between real analysis (PM V) and the theory of real numbers (PM VI) is, in PM, ultimately grounded on the fact that the two doctrines are used in two very different ways.

Why should one first not restrict the definitions of limit, convergence and continuity to real valued functions? Because there is a context, where real valued functions do not occur, and where limits, convergence and continuity play a crucial role. In mechanics, these notions are routinely used to study the properties of non-numerical functions, which take as arguments instants of time and as values points in space. It is for this reason that one should generalize the definitions of the basic notions of the differential calculus so that they could be applied to non-numerical functions. Russell (1919) is very clear on this point. After having resumed the standard numerical definition of continuity, he explains (p. 114):

> We may now generalize our definitions so as to apply to series which are not numerical or known to be numerically measurable. The case of motion is a convenient one to bear in mind ... It is the meaning of 'continuity' involved in [statements like 'all motions are continuous'] which we now wish to define as simply as possible.

After having summarized the ordinal definition of limit, convergence and continuity expounded in PM V, section C, he adds (p. 116): 'there is thus nothing, in the notions of the limit of a function or the continuity of a function, that essentially involves number'. In other words, the ordinal account is better than the standard numerical approach because it supplies a simple explanation of the application of real analysis in mechanics, where the arguments and the values of functions considered are instants and spatial points, and not numbers. Note that this idea played a crucial role in Russell and Whitehead's attempts to construct the real world. In Whitehead's *The Concept of Nature* (1920, pp. 49–73), as in Russell's *Our Knowledge of the External World* (1914b, pp. 120–34), the nested interval technique (and thus, the notion of convergence) is directly applied to the series of 'events'. As the idea is to define the fundamental concepts of mechanics (the notion of an instant in the case considered) from the rough data of experience, one cannot rely on a prior operation of measurement.[26] Whitehead (1920) states the point in a very vivid way. Thus, after having presented his construction, he remarks (p. 65): 'nothing has been said as to the measurement of time, [since] such measurement does not follow from the serial property of time; it requires a theory of congruence which will be considered in a later lecture'. The issue concerning the application of real analysis is then crucial, since the restriction of real analysis to the theory of real valued functions would prevent Russell and Whitehead from developing their constructionalist programme.

The need to account for the application of limit, convergence and continuity in mechanics explains why one should espouse the ordinal conception of real analysis. But I have not yet said anything about real numbers. Real analysis gives us no reason to introduce them. Why then seek to define them? Russell and Whitehead justified their decision by referring to the particular application of numbers in measurement. Numbers are not connected to the foundations of real analysis; but one needs a theory of rational and real numbers for explaining measurement – more particularly, for accounting for the introduction of coordinates in geometry. As we have seen in Chapter 5, Russell and Whitehead's doctrine of measurement is nothing but an abstract characterization of the coordination process, which played a very important role at the time. The key concept of the whole construction was the notion of a Möbius net, which allows for the introduction of rational coordinates. In Russell and Whitehead's doctrine, this emphasis on the rational case is taken up again. Measurement is first defined for rational numbers and then extended to real numbers. Of course, to achieve this extension, Russell and Whitehead had to introduce considerations about continuity and limit of the vector-families. But, as the stress put on the rational case shows, the special features which make the real numbers worth studying have nothing to do with ordinal considerations. Real numbers must be defined because the relations (that they are) measure (when their fields are suitably restricted) the relations which they connect. The consideration of the applications is not here only a reason to favour one definition of reals among many possible alternative ones – it first and foremost justifies the need to have a definition of the real numbers. This stands in sharp contrast with what happened in PoM, where the arithmetical definition of real numbers appeared to be completely useless.[27]

Let me come back to my initial question. At the beginning of the twentieth century, real analysis and the theory of real numbers were often seen as two components of a unified theory, the theory of real valued functions. What is distinctive of PM is that the unity of this field is broken up into two subdisciplines: the ordinal theory of real analysis and the relational conception of real numbers. What justified this move? Nothing other than a reference to applications. Russell and Whitehead's reasoning was the following. Concepts like limit, convergence and continuity are, in mechanics, applied to non-numerical functions, whereas the notion of real numbers is used to measure quantity (to introduce coordinates). According to *Applic*, this difference between contexts of use should be reflected in the definitions of the concepts considered; one should thus favour an ordinal account of real analysis and a relational theory of real numbers, which both account for the different uses of these concepts.

What is the benefit of this long detour in PM V? In the previous chapter, devoted to PM VI, I proceeded as if there were no question about the need to develop a theory of real and rational numbers. The only issue there was to arbitrate between different possible theories of the rationals and reals. We can now see that this picture was far too simple. The real problem, for Russell and Whitehead, was not to single out a definition of reals among many possible ones. It was to justify the need to develop a theory of real numbers in the first place – that is, to justify the reorganization induced by the splitting of the theory of real valued function into two separate fields. *Applic* in PM is instrumental in the defence of the idea that the theory of real numbers represents an important branch of the mathematical sciences. As I said at the beginning of this chapter, a standard way of understanding *Applic* is to decompose it into two distinct theses: the claim that there is a distinction between a given mathematical concept and its concrete applications (the abstract theory of real numbers and its concrete applications in measurement) and the claim that the connection between the concept and its applications should nevertheless be made in the definition. In this model, *Applic* is seen as a principle which relates two separate domains. For instance, Whitehead's metaphysical claim, that the more abstract a mathematical concept is, the more concrete its applications are, complies with this pattern. But the PM use of *Applic* does not conform to this schema. The scientific architecture (i.e. the fact that a given field of knowledge is separated from another) is not in PM considered as given, prior to and independent of *Applic*. What is at stake is not whether or not a given already defined mathematical concept should be related to its external application; the genuine question is rather to fix the outlines of the concept whose applications are considered. In a certain respect, this is strange: one could indeed consider that, in order to be able to look at the applications, one must already know 'what' it is that is supposed to be applied. But our previous analysis shows how the consideration of the applications can help to determine the identity of what is applied. Thus, Russell and Whitehead show that the theory of real valued functions is applied to two very different contexts, and they claim that this heterogeneity should be taken into account in our definitions of 'what' is applied. In the PM use of *Applic*, the identity of the 'what' is thus not known beforehand; on the contrary, it is by looking at the various contexts of use that one can see whether a given mathematical notion should be construed as an independent concept, or whether it should be integrated within a larger body. *Applic* in PM has thus some essential relation to the 'architectonic' question of how to carve out the mathematical content.[28]

This feature is related to another one. In PM, the applications considered are not necessarily extra-mathematical. Thus, as we saw in PM VI, measurement is first and foremost a generalization of the geometrical process of coordination of the real projective line; and, as mechanics is considered in PoM as a part of mathematics, the reference to the application of real analysis to mechanics could also be regarded as an example of intra-mathematical application. When Russell and Whitehead spoke about the application of a mathematical notion, they usually simply referred to a context which was not the one that we usually think of when we consider that notion. This did not necessarily imply that the context considered was extra-mathematical. Thus, Dedekind for instance dealt with real numbers in the course of research on the foundation of the differential calculus, and we saw that many mathematicians, at the time, followed Dedekind in connecting real numbers with real analysis. Now, Russell and Whitehead considered that this view on the global architecture of mathematics was wrong, and that real numbers should be linked to their use in the introduction of coordinate in geometry. At the time, it was unusual to lay so much stress on this context of use, and this is the reason why 'measurement' was said to be an application of numbers. In PM, the term 'application of a mathematical concept' did not mean 'extra-mathematical use of a mathematical concept'; it only meant 'unexpected use of a mathematical concept'. One finds the same idea again: *Applic* in PM is a principle which helps us to carve out the mathematical content; it is not a thesis about the relation between the mathematical (abstract) and empirical (concrete) sciences. It is this 'architectonic use' of *Applic* that I will continue to study by focusing, in the following, on the recent discussion about the legitimacy of *Applic*.

6.4   *Applic* in neo-logicism

As is well known, neo-logicists claim that mathematics (arithmetic, real analysis and set theory) can be derived from second-order logic, supplemented by some suitable abstraction principles.[29] The debate opened by this programme usually revolves around the nature of abstraction principles (are they definitions, logical principles, etc.?) and around the way one can differentiate 'safe' abstraction principles from 'bad' ones (which leads to a contradiction).[30] The question I will focus on is different. Let it be granted that abstraction principles are acceptable from a logicist point of view; now, in many cases, one can use different 'safe' abstraction principles to get what one wants. How can we decide which path is the best? This is the problem Wright faces in his 'Neo-Fregean Foundations for Real Analysis' (2000).

In Chapter 5, I presented Frege-Hale's definition of real numbers as the ratio of quantities. Wright expounds an alternative characterization brought forward by Shapiro (2000), which essentially amounts to rewriting Dedekind's definition in abstractionist terms.[31] Wright shows that the two constructions are irreproachable from a logical point of view and seeks to determine which development is the best. In the course of his argument, Wright (2000, pp. 319–20) draws a comparison with the definition of natural number. In arithmetic too, two distinct processes can be used: one can get the natural numbers in what has become the standard way, by making use of Hume's Principle; but one can also maintain the path first explored by Boolos (1986), consisting in modifying Frege's axiom V so as to get a consistent abstraction principle, called 'new V', which allows us to develop, within the abstractionist setting, a sufficient amount of set-theory to obtain arithmetic. Here as well, says Wright, one has to choose between two logically correct ways of developing the neo-logicist programme.[32] But how are we supposed to decide which path to take?

Shapiro, Hale and Frege do not answer this question in the same way. Before describing the nature of their disagreement, let me insist on their shared assumptions. The three philosophers agree that any philosophical account of numbers must endow them with their usual mathematical properties. This idea is encapsulated in Shapiro's 'faithfulness constraint', hereafter written as *Faithful* (2006, p. 111):

> As I see it, the goal of philosophy of mathematics is to *interpret* mathematics, and articulate its place in the overall intellectual enterprise. One desideratum is to have an interpretation that takes as much as possible of what mathematicians say about their subject as literally true, understood at or near face value. Call this the *faithfulness* constraint. According to *ante rem* structuralism, natural numbers are places in structures, and places in structures are *bona fide* objects. This accords with faithfulness. Grammatically, numerals seem to function as singular terms, and according to *ante rem* structuralism, numerals are singular terms.

To be acceptable, a philosophical theory of natural numbers and real numbers should be faithful, that is, it should grant them the properties mathematicians give them. Thus any account of the whole numbers should construe them as elements of an ω-sequence, and any account of the real numbers should construe them as members of a dense and complete ordered set. *Faithful* is, for Shapiro, a minimal constraint: any unfaithful account would be discarded out of hand, since it will not be an account of the mathematical structure one seeks to capture. Hale and Wright accept this condition. The import of the so-called 'Frege's theorem' is thus to show that, from second-order logic and Hume's abstraction principle, one can derive Peano's postulates. Note, however, that, despite its apparent innocuousness, *Faithful* commits us to the view that one already knows, before the analysis takes place, what the mathematical content is that one wants to capture. In this respect, it conflicts with the position I attributed to Russell and Whitehead above – but more on this later.

However, Shapiro, Hale and Wright believe that satisfying *Faithful* is not enough. If any ω-sequence can provide a foundation to arithmetic, it is not true that $\mathbb{N}$ may be characterized as any ω-sequence. And the same holds for the real numbers. Even if any structure isomorphic to $\mathbb{R}$ can provide a foundation to the theory of real numbers, it is not true that the reals should not be identified with any dense and complete ordered set. From 'the *epistemological* project of providing a foundation for [the] standard mathematical theory of [the objects in a given field of mathematical enquiry]', Wright claims that one should distinguish 'the *metaphysical* project of explaining the nature of [those objects]' (2000, p. 320). That the natural numbers, as defined via new V or Hume's Principle, satisfy Peano's axioms shows that the two accounts fulfil the first task – but not that the two accounts fare equally well on the second one. It is on this point that the three philosophers disagree.

Shapiro (2006, p. 112) complements *Faithful* with another principle, called the 'minimalism constraint', according to which philosophers should 'not attribute *mathematical* properties to mathematical objects unless those attributions are explicit or at least implicit in mathematics itself'.[33] It is not only what mathematicians say, but also what they do not say, that a good philosophy of mathematics must take account of. Philosophers should not attempt to fill the gaps left by mathematicians, and they should therefore not attempt to define numbers in terms of other objects (like sets or equivalence conditions).[34] This 'minimalism constraint' is of course tailored to meet the *ante rem* structuralist requirement: the only possible way to satisfy the constraint is to define a whole number as a place in an ω-sequence and a real number as a place in a continuous series.[35] For Shapiro then, no abstractionist construction satisfies the minimalism constraint and, henceforth, all of them should be discarded as metaphysically incorrect.

For Hale as for Wright on the other hand, *Applic* is a metaphysical principle enabling us to single out the good definition from among the many epistemologically correct ones. Thus Wright explains (2000, p. 325):

> [*Applic*] and the insistence on a contrast between establishing a mathematical theory and merely establishing a theory which allows of interpretation as that theory have in common the thought that the objects of, for example, the classical theories of the natural and real numbers ... have an essence which transcends whatever is shared by the respective types of models of even categorical (second-order) formulations of those theories. [*Applic*] explicitly incorporates the additional thought that this essence is to be located in the applications ... [The distinction] between deriving the axioms of number theory or analysis and merely deriving a body of statements which allow of interpretation as those axioms ... might in principle ... be grounded in some other kind of conception of what makes for the essence of natural or real number. But no candidate is on the table besides that incorporated in [*Applic*]. And it is hard to see what alternative there could be. For the pure mathematical theories of those entities make no distinction between them and any other isomorphic structure – so what could distinguish them except something to do with application?

Since 'the pure mathematical theories' make no distinction between the structure of real numbers (for instance) and 'any other isomorphic structure', it is by looking at the applications that one can uncover the essence of the real numbers and distinguish the right structure from their isomorphic rivals. For Hale, then, the account of natural numbers via Hume's axiom and the account of real numbers via the axiomatization of q-domains are preferable over the alternative approaches, because they both satisfy *Applic*.

Wright does not endorse Frege-Hale's definition of the real numbers.[36] Indeed, the parallel he draws between the natural and the real numbers is not intended to support the idea that what holds in the first case holds in the second, but, on the contrary, to bring forward the specificities of the arithmetical case.[37] His idea is that, in the case of natural numbers, the knowledge one has is 'induced by reflection upon sample, or schematic, applications' (2000, p. 327), whereas 'there is simply no such thing as determining a *real* value of a quantity by measurement or indeed by any other empirical procedure' (2000, p. 328). Wright even suggests that, in the case of real numbers, 'the flow of concept-formation goes *in the other direction*': instead of being extracted from its applications by reflection, 'the classical mathematics of continuity is made to inform a *nonempirical* reconceptualization of the parameters of potential variation in the empirical domains to which it is applied' (2000, pp. 328–9). Wright makes then the justification of *Applic* dependent upon the epistemological nature of the mathematical knowledge considered. When elementary mathematical knowledge is considered (elementary arithmetic, elementary geometry), *Applic* should be taken into account – but in more advanced domains, the structuralist point of view must be adopted. Wright attempts thus to reconcile the two opposite standpoints of Hale and Shapiro by relativizing the use of *Applic* to the nature of the mathematical field considered. In Table 6.1 I summarize the positions of the three philosophers.

From our perspective, what is important is not the differences between the three conceptions, but their common core. The three philosophers consider that *Faithful* should always be followed and that *Applic* is a principle allowing us to answer a metaphysical question concerning the nature of numbers. These two assumptions are linked together: *Faithful* guarantees that the identity of our mathematical objects, qua mathematical objects, is already fixed – and thus, that the sole remaining question is to know how one should characterize in a general way the nature of these entities. The adoption of *Faithful* leads us to construe *Applic* as a metaphysical principle, whose aim is to display the metaphysical nature of the mathematical objects – though not as an epistemological one, whose goal would be to identify the mathematical objects. For Hale, Wright and Shapiro, this second task is within the competence of mathematicians – and it is perfectly solved by them.

This division of labour is of course not the one we encountered in PM. In the previous section, I argued that *Applic* is used in PM as an architectonic principle, whose aim is to explore the different possible ways of articulating the mathematical material. In PM, *Applic* is thus not used as a tool enabling us to meet a metaphysical issue; it is an instrument allowing us to solve the 'epistemological' (I say 'architectonic') problem regarding the identification of the mathematical objects. What makes Russell and Whitehead's use of *Applic* specific is the fact that they do not combine *Applic* with *Faithful*. Prior to and independently of *Applic*, there is no way in PM to guarantee that the mathematical material is compartmentalized in the right way.

Since it is essential for the comparing, let me come back to *Faithful*. The principle demands that one takes 'as much as possible of what mathematicians say about their subject as literally true'. According to *Faithful* then, to identify the subject-matter of any mathematical theory, one should just look at the syntax of its theorems. Natural numbers occur in arithmetical

Table 6.1 Summary of the different positions of Hale, Shapiro and Wright

	Faithful	Minimalism	Applic
Shapiro	Yes	Yes	No
Hale	Yes	No	Yes
Wright	Yes	Depends	Depends

statements, so arithmetic is about natural numbers. Real numbers occur in propositions from real analysis, so real analysis is about real numbers.[38] In this conception, the surface form of the mathematical discourse gives us enough information to divide the mathematical knowledge into different topics. The idea that one can learn from the mathematicians how the various mathematical fields are defined and articulated is then a direct consequence of the idea that there is no gap in mathematics between the real and the apparent form of the propositions. This is of course something that Russell and Whitehead forcefully denied in PM. In the next section, I will focus on this rejection and on the consequences it has on the use of *Applic*.

6.5   *Applic* and the distinction between logical and apparent forms of propositions

In Chapter 3, we saw that, in PoM, Russell's analysis should be viewed as a progressive adjustment between the actual state of existing mathematics and the logical framework. We particularly insisted on the fact that the information flow goes in both directions: existing mathematics pulls the logical development but, at the same time, the organization of pre-logicized mathematics is revised so as to fit easily the relational mould. When dealing with Russell's programme, two symmetrically opposite mistakes should then be avoided: (1) we should not claim that Russell wants to represent faithfully in his logical system the existing mathematics exactly as it is; (2) we should not claim that Russell extracted the organization of the mathematical content from the sole relational framework. This is the context which should be kept in mind when examining the role of *Applic* in PM. The situation there was roughly the same as in PoM. The mathematics Russell and Whitehead wanted to analyse was the mathematics as it was practised in those days by the working mathematicians, though they did not slavishly follow their customs.

The adoption of *Faithful* seems to lead us into making one of the two opposite mistakes – that is, to believe that the division of the mathematical field is already given, and that the sole reasonable attitude in philosophy of mathematics is to follow the way mathematicians carve out their material. On the other hand, Russell and Whitehead's rejection of *Faithful* seems to carry us into the vicinity of the other mistake. If the logical form has nothing to do with the surface grammar of the mathematical propositions, then it seems that the organization presented in PM is nothing but the projection of the logical scaffolding into existing mathematics. How then can we avoid these two symmetrical pitfalls? Before examining the question in more detail, let me settle a side issue.

In the literature, there is no consensus on whether Russell espoused the dichotomy of apparent versus real form before or after 'On Denoting'

(1905c).[39] Hylton (1990) considers this essay to represent a radical break: before 1905, Russell would have espoused *Faithful* and regarded grammar as a reliable 'guide' (PoM, p. 42); after 1905, grammar was considered the main obstacle to analysis. Against this interpretation, Levine (2002; 2009) rightly points out that there are many examples in PoM where the real form of the proposition does not match the grammatical pattern. Levine's favourite illustration is the analysis of cardinal numbers. Russell's Pieri-inspired definition of projective order in terms of incidence is another beautiful example. Let me recall that 'the points a and b are separated from the points c and d', whose apparent form is '$R((a, b), (c, d))$', is translated into 'there are two points x and y such that a and b, and c and d, are harmonic conjugates with respect to them', whose form is '$\exists x \exists y(H((a, b), (x, y)) \,\&\, H((c, d), (x, y)))$'. The grammatical form of the sentence is completely upended in the process. More generally, all the analyses I have developed in Chapters 1, 2 and 4 lead me to subscribe to Levine's standpoint. The theory of incomplete symbols developed in (1905c) makes of course a difference – but this difference is of degree, not of kind. Russell in PoM did not endorse *Faithful*: the idea that analysing is restructuring pre-dates 1905.

Let me come back to my main line of interest. Today, the standard argument against the rejection of *Faithful* takes the following shape. If the surface forms of the mathematical statements do not match their deep logical form, then one would have to explain how we could have access to the real constituents of the mathematical proposition. But how can we explain this? It seems that one could not rely on any extra-linguistic capacity to scan, through the veil of the language, the structure of the propositional content. Owing to this, it is preferable to keep close to the grammatical structure of the mathematical statements. In other words, any departure from *Faithful* should be justified by a story explaining how, in the case in point, one is in a position to tell what the constituents of the targeted sentences are. Now, when mathematics is in focus, no such a story is available – it seems difficult to hold that one has an extra-linguistic perceptual-like access to the inner composition of number. Benacerraf (1965) was the first to have articulated this defence of *Faithful*. He took the well-known example of Zermelo and von Neumann's definitions of the finite ordinals: 'there is no way connected with the reference of number words that will allow us to choose among [these two set-theoretical rewritings], *for the accounts differ at places where there is no connection whatever between features of the accounts and our uses of the words in question*' (1965, p. 62). Conclusion: either one accepts the existence of a non-linguistic access to the real constituents of the propositions, or one sticks to *Faithful*. One cannot however reject both, on pain of being unable to choose among the multiple possible equivalent definitions. Moreover, as it is difficult in mathematics to defend the first horn of the dilemma (we seem to have no intuitive access to abstract entities like mathematical objects), one should stick to *Faithful*.

To a certain extent, this presentation fits well with what we find in Russell. In particular, there are some passages where Russell seemed to claim that one always has extra-linguistic access to the real constituents of the propositions (1912a, pp. 51–2):

> The fundamental principle in the analysis of propositions containing descriptions is this: *Every proposition which we can understand must be composed wholly of constituents with which we are acquainted* ... It is scarcely conceivable that we can make a judgment or entertain a supposition without knowing what it is that we are judging or supposing about. We must attach *some* meaning to the words we use, if we are to speak significantly and not utter mere noise; and the meaning we attach to our words must be something with which we are acquainted.

Acquaintance (in 1903 as in 1912) designates an immediate cognitive relation between our mind and a non-mental entity – a sort of direct intuition moulded from sense perception. In the (notorious) passage quoted above, Russell contends that one has an extra-linguistic access to the constituents of the propositions one understands – and thus that one has an acquaintance with the ultimate constituents of mathematics. In accordance with the Benacerrafian schema, the 'principle of acquaintance' would then justify the distinction between real and apparent form in mathematics. A non-linguistic intuition would provide us with the means to single out the good analysis from among the many possible ones. This reading does not bear scrutiny, however. Russell never relied on acquaintance for identifying the indefinables of mathematics. As he explained, 'it is often easier to know that there must be such entities [the indefinables] than actually to perceive them' (PoM, p. xx). More to the point, in PoM, as in PM, acquaintance is never referred to for justifying the different analysis. For instance, Russell does not justify his analysis of projective order by an appeal to our acquaintance with the ultimate projective relations. In PM, Russell and Whitehead never said that their relational recasting of the quantitative statements was based on an acquaintance with the real constituents of quantity. Russell's rejection of *Faithful* cannot then be grounded on the acquaintance principle, since this principle never played any role in Russell's actual analysis.[40] But then, how can we avoid the multiple reduction problems?

As we found in Chapter 3, we have to ask: what are the criteria Russell used to delineate the real form of the mathematical content he was considering? In PoM, Russell did not derive the articulation of mathematics from logic alone, but, at the same time, he did not merely ratify the mathematical architecture as he found it in the mathematical textbooks. The same is true in PM: the distinction between real analysis and the theory of real numbers was not usually made in the discourse of the mathematicians (in the textbooks); but, for all that, this distinction was not based on a direct acquaintance with the hidden logical forms. How then are we supposed to understand the process by which the real structure of the proposition is brought out of its linguistic gangue? The all-or-nothing dilemma in which Benacerraf puts us clearly makes Russell's analytical process unintelligible. For Benacerraf, one can either espouse *Faithful* or postulate an extra-linguistic access to the content. Now, if one assumes *Faithful*, then the analysis terminates before it begins – since the given syntactic form of the sentence is viewed as a transparent window through which the structure of the mathematical content can be seen. But if one rejects *Faithful* and postulates an immediate cognitive access to the content, then we end up in the same situation: understanding a sentence will automatically give us knowledge of the ultimate constituents of the propositions considered (since, now, understanding amounts to having an acquaintance with the constituents). The difficulty comes from the fact that the 'logic' governing the use of the concept of an 'access' is binary: either we have access or we do not. Now, Russell's notion of analysis requires more articulation. What is analysed must be understood, otherwise analysis could not even start; but, at the same time, one should be able to understand something without being able to analyse it, otherwise analysis would never be a task. If Benacerraf's all-or-nothing alternative were true, then Russell and Whitehead's position would be unintelligible. But are we compelled to endorse Benacerraf's view?

It indeed seems that what Benacerraf's dilemma prohibits is what mathematicians themselves never stop doing. Mathematicians do not often take mathematical language at face value; at the same time, they still continue to claim that certain definitions are better than others, without ever grounding their claim to any special cognitive acquaintance with the ultimate constituents of the mathematical content. Take the example of the definitions of projective space. I explained in Chapters 1 and 2 that many different concurrent conceptions coexisted during the nineteenth century. Some people, following Plücker and Klein, defined the notion in an analytic way, *via* the homogeneous coordinates; others, following Poncelet and von Staudt, tried to characterize it in a synthetic way. How is it possible to take mathematical language at face value, in this case, where two formulations of the same fundamental concept coexist? This situation is not an exception – history of mathematics is replete with cases where mathematicians disagree on the ways of presenting a concept or a theory. To settle this kind of dispute, mathematicians have never resorted to an extra-linguistic intuition: the followers of Plücker never claimed to have an acquaintance with the *n*-tuple of numbers that were the projective points; and the partisans of Poncelet never said that they knew by intuition that projective points were not numerical. Benacerraf's idea that one should either stick to the surface form of the mathematical sentences or rely on an extra-linguistic cognitive relation to the mathematical content is thus too coarse: it prevents us from understanding what mathematicians do.

My suggestion is that the use of *Applic* in PM should be understood as an extension of the way mathematicians settle disagreements of this kind concerning the definitions of their basic concepts. To discriminate between the different competing analyses, mathematicians attempt to find a certain context in which the concept considered can be more easily used when framed in a certain way. Of course, the 'discriminating' context should not be the standard one, since the different versions roughly perform in the same way in this case. Once it is established that a certain definition of a concept explains in a very natural way its application in the relevant context, the partisans of this approach usually try to convince their colleagues that this context is especially important – so important that any definition which does not account for this use should be disregarded. To come back to our earlier example, the followers of Plücker insisted on the fact that projective space (complex projective space) was the right setting to classify algebraic curves and algebraic surfaces.[41] They went on by noting that the synthetic approach did not allow us to relate directly the concept of projective space to the theory of algebraic curves. And they ended by claiming that developing a correct classification of curves and surfaces is, at least since Newton, one of the more important issues in mathematics. The conclusion that followed was that one should resolutely espouse the analytic definition of projective space and make the most of the homogeneous coordinates, since only this approach enabled us to do important things in mathematics. On the other hand, the late followers of Poncelet insisted on the fact that projective geometry provided models for the various classical non-Euclidean theories, and then unified a field of research, i.e. geometry, whose history can be traced back to Euclid. To adopt the analytical definitions would amount to diluting geometry in algebra to the extent that its identity would be lost, and therefore to miss the fact that projective geometry gives to the whole field its unity and coherence. The next conclusion that followed was that one should resolutely adopt the synthetic definition of projective space, since it is the only one which can restore the unity to geometry after the emergence of non-Euclidean theories. In both cases, the way one applies a concept teaches us something about the way one should define it – as an algebraic manifold in one case, as an incidence structure (if one espouses Pieri's approach) in the other. Contrary to what Benacerraf suggested, mathematicians are then not powerless when confronted with a choice between alternative definitions. What they usually do is to search a context of use which can discriminate between the candidates in contention.

I think Russell and Whitehead's understanding of *Applic* in PM follows this pattern. The idea is that, by varying the contexts of use of the concept considered, one gains a better view of its internal constitution. Real numbers, limit, continuity and convergence are notions that are easily amalgamated when one restricts one's attention to the real valued functions. As soon as some other contexts are considered, however, like mechanics and measurement, certain differences, not perceptible at first, emerge, which are not taken into consideration in the usual characterizations. Now, one could well stick to the standard definitions and claim that these contexts are not the most important ones. But one can also adopt a different stance, and build these distinctions in the definitions themselves. When disagreements about definitions occur in mathematics, there is no other solution than to find a context that could act as a test case differentiating the various competing approaches of a given notion. Language is here not taken at face value (since the disagreement is precisely about language), and no reference is made to a direct cognitive access of the content. It is the confrontation of the various definitions with the different contexts of use which enables us to weigh up the different solutions against each other. *Applic* in PM is nothing but the extension of this very common strategy.

The parallel between *Applic* in PM and what mathematicians never stop doing in their work is not intended to provide a full justification of *Applic*. The point of the comparison is to deepen the gap between PM and the contemporary use of *Applic*. As we have seen in the last section, Shapiro, Hale and Wright, despite their differences, all agree on the fact that *Applic* must be construed as a 'metaphysical' principle and not as an 'epistemological' one. The idea is roughly that mathematics provides philosophers with some well-defined structures, and the only remaining job is to formulate principles (like *Applic*) allowing us to choose, among the many isomorphic alternatives, the one which is consonant with the essence of the concept. In PM, no such division of labour between mathematics and metaphysics existed. For Russell and Whitehead, it was just not true that mathematicians agree on the identity of their objects. Sometimes, they do; but not always. And *Applic* was precisely used in PM when no agreement existed. Thus, *Applic* in PM VI was not only a means for discriminating between different definitions of real numbers; it was used to show why one cannot do without real numbers in mathematics, even though one can do without real numbers in real analysis. In other words, Shapiro, Hale and Wright seem to view mathematics as a very peaceful place – a place where it is always possible to come to an agreement concerning questions like which mathematical concepts are the most important ones? For those philosophers, there are no philosophical problems in mathematics; the difficulties only begin when mathematics is left out and general ontology is reached. Russell and Whitehead did not share this conception. For them, mathematics was not a quiet place. Mathematicians often do not agree on the way their science should be articulated (on the way one should answer such questions as: Which theories are the most fundamental ones? Which presentation of a certain theory is the best one? What is the best proof of a given theorem?

Etc.), and they seek to defend a certain view against alternative ones. Russell and Whitehead saw their own reflection as a continuation of these discussions.

This contrast between PM's architectonic use of *Applic* and its contemporary understanding explains also why there can be no answer to the question raised at the beginning of this chapter: how did Russell and Whitehead justify *Applic*? The fact is that one cannot not use *Applic*. To defend a particular definition, and to defend also the need for this definition, there is no other way than to find a discriminating context. The real question then is not whether one should use *Applic* – the real issue is how to use *Applic* in a convincing way. *Applic* says that the definition of a concept should account for its application; it does not say what the applications one should account for are. It is at this stage that disagreements occur. Thus the decision to relate real numbers to measurement is questionable – one could refuse (with Dedekind for instance) to consider that this is an important application, and argue that $\mathbb{R}$ is only used to ground the differential calculus. The choice to downplay the importance of the theory of the real valued function could equally be challenged – since this theory was instrumental in the development of Lebesgue's integral and was one of the sources of the functional analysis.

What are the general criteria which enable us to determine which contexts of use are important, and which are not? In PM, reasons are each time given for the way *Applic* is in fact used. But it is important to understand that these reasons always have a local character; they are not grounded on a general theory. That is, *Applic* is not a recipe, which, once adopted, mechanically delivers a certain result. The way one implements *Applic* in particular cases is more important than the decision to use it. One could always downplay the importance of the discriminating context – one could also challenge the way one articulates it to the other ones. The weak point of the contemporary discussion is that it only revolves around the legitimacy of *Applic*. What really matters is the way one implements it in particular cases.

6.6   Conclusion

I began this chapter by asking: Why did Russell and Whitehead adhere to *Applic*? The answer I gave was that this is not the right question to ask. As I have explained, the metaphysical interpretation assumes that there is a consensus in the mathematical community on the way one should carve out mathematics – a consensus, for instance, on the fact that the concept of real numbers plays a fundamental role and that, therefore, any account of mathematics must, at some stage, meet this notion. Russell and Whitehead did not see the situation that way. For them, the architecture of mathematics was not given in the existing mathematical practice: mathematicians often disagree on the way they should frame their basic notions; they even disagree on whether a given concept (real numbers, for instance) is fundamental in mathematics or not. The usual way to settle these kinds of dispute is to find a 'discriminating' context and show that the use of the concept in this context plays a crucial role in the mathematical sciences. PM's use of *Applic* is nothing but an extension of this usual strategy.

The interpretation developed here is then in line with the developments presented in Chapter 3. There, I showed that Russell never gave any criteria allowing the drawing of the demarcation between the logical content of a mathematical theory and the psychological elements misleadingly associated with it. Considerations from logic and from existing mathematics were each time supplied to determine the course of the dividing line in particular cases; but the balance of the two ingredients was not, in PoM, fixed once and for all. One finds a parallel situation in PM. The logical form of the propositions did not resemble the grammatical structure of the sentences of existing mathematics; and yet, no non-linguistic access to these real forms was provided. As we have seen, *Applic* was a means to explore and weigh up the different possible analyses, but the guideline was very vague and could be applied in many different ways. *Applic* was more a guiding rule of thumb than a recipe that mechanically gave each time it is used a definite result. In PM as in PoM, the determination of the logical structure of the content remained a matter of combining and adjusting heterogeneous pieces of data that should be assessed only on a case-by-case basis.

To conclude, I would like to come back to the comparison between Russell and Whitehead's use of *Applic* and Frege-Hale's, made in Chapter 5. My main contention was there that the way Russell and Whitehead implemented *Applic* was, conceptually speaking, more satisfying than the way Frege and Hale did. At the beginning of this chapter, however, I suggested that another question was more pressing: is it legitimate to use *Applic*? We now understand that this last move did not go in the right direction. *Applic* was not in PM the metaphysically loaded principle that it has become in neo-logicism. The main difficulty for Russell and Whitehead was not to justify the use of *Applic*, but to use it in a 'convincing' way. In this respect, one can only acknowledge the success of PM VI. Let us recall that, contrary to what happened in Frege-Hale's doctrine, the arithmetical properties of numbers were not, in PM, directly inherited from the properties of the q-domain; and conversely, the notion of magnitude was not restricted to that of extensive magnitude. In other words, in PM, the connection between number and magnitude was flexible enough to allow the elaboration of a theory of pure numbers (section A) and of a theory of magnitude (section B) – each one independent from the theory of measurement (section C) and from each other. Russell and Whitehead succeeded in integrating into one coherent whole several very heterogeneous requirements (the idea that numbers should be linked to measurement, the claim that numbers are logical objects, the idea that quantity should not be restricted to extensive magnitude). This is the way *Applic* was implemented, not the fact that *Applic* was endorsed, that makes the PM construction philosophically so interesting and powerful. My aim in this chapter was to expound and defend this shift in the perspective.

7
Russell's Universalism and Topic-Specificity

A tension runs across all the six preceding chapters. On one hand, we have claimed that Russell wanted to provide his readers with more than a formally perfect substitute for the mathematical concept he considered; on the other hand, we have insisted on the fact that he never expound the additional criteria he used to select his favourite analysis from among the many possible ones. Russell's decisions were not arbitrary; in each case, reasons were provided to justify the choices. But these justifications were always tied up to some local and topic-specific considerations, and they were not grounded on any general criteria. Thus, we saw in the last chapter that *Applic*, one of the best candidates to play the role of a criterion, was no more than a guiding rule of thumb. In other words, in the picture I have painted of PoM and PM so far, there is a maladjustment between Russell's general characterization of logical analysis (as a formally perfect, or truth-preserving, translation of the mathematical theorems into the logical language) and his own practice of analysis (which takes into account non-formal aspects of the mathematics he analysed).

In Chapters 3 and 6, I focused on two ways of expounding this difficulty. I first showed in Chapter 3 that Russell's sensitiveness to the topic-specific features of mathematical practices could become the prey of his anti-psychologism. In his reply to Boutroux, Russell explained that some characteristics usually associated with certain mathematical concepts have in reality nothing to do with their mathematical content. The argument could easily be turned against Russell himself – against his own analyses of projective and metrical geometry in PoM, and of rational and real numbers in PM. In Chapter 6, I tackled the same problem from a different angle, focusing this time on the distinction between the apparent and real form of mathematical propositions. Russell claimed that the logical structure of the mathematical theorems does not necessarily follow the grammatical pattern of the theorems. But he did not believe that logicians have a direct access to the real form of propositions, and he did not give any general criteria enabling us to single it out.

These two difficulties are in fact the two facets of the same problem: what is the nature of the relationships between what Russell called 'existing mathematics' and the mathematics developed within the logical system? For Russell (as for Frege), logic is universal: it applies to everything and does not take into account any non-logical (physical, biological, metaphysical, etc.) distinctions. Logic applies to all fields of knowledge equally; and no restriction over the domain of the variables is logically justified. But if so, then any distinction between mathematical domains or topics should be ignored in logic. Yet, we have seen in the previous chapters that, in order to develop the logical system, Russell heavily relied on some facts concerning the historical development of mathematics. The differences between topics, as they emerge in the practice of the mathematicians, do not count for nothing in PM or PoM. How then can we reconcile Russell's universalism with the idea of topic-specificity? How can one consider logic as the first science, while relying on an extra-logical source of knowledge (the knowledge of existing mathematics) for developing the logical system?

In this concluding chapter, I will attempt to show that Russell's sensitiveness to the non-formal features of mathematical knowledge does not go against his logical universalism. My idea, in short, is to use a distinction between two kinds of universalism introduced by François Rivenc (1993). I will explain more fully the distinction in what follows, but the key feature of the sort of universalism ascribed to Russell is that logicism is conceived as a work in progress, not as an already completed system. As a consequence, in this framework, one should not press too much on the separation between what is externally given to the logicist and what is logically reconstructed. There is a great flexibility about these matters, which is due to the fact that what is regarded as external to the logical system at a certain stage can become a part of the system at a later stage. All the developments which I have spoken about in the previous chapters have an intermediate status: they were preparatory investigations which aimed at fitting the historical data into the logical setting, and which, at the same time, probed the capacity of the logical machinery to capture the mathematical material. All these discussions were then both inside and outside the system: they were places where what was considered as external to the relational framework was on the point of becoming a part of it. This distinction and the stress put on progressivity and time will shed some light on the peculiar way Russell related logic and the existing mathematics.

In a first section, I will expound in general terms the difficulty encountered in the previous chapters, and present two non-Russellian ways of solving it – one inspired by Frege, the other by the later Wittgenstein. In section 7.2, I will use the distinction between positive and negative universalism made by Rivenc (1993) to expound what is peculiar to Russell's approach and which makes it distinct both from Frege's and Wittgenstein's solution. In section 7.3, I will substantiate this analysis by focusing my attention on the notion of a mathematical content. In section 7.4, I will make some remarks about the possible impact of Russell's project (as understood here) on contemporary philosophy of mathematics. I will hold that the importance given by Russell to the contingent historical considerations forbids any attempt to revive his logicism in its original form. But I will nevertheless claim that his logicism compels us to challenge seriously the opposition made today between 'foundationalist' and 'practice-based' approaches in philosophy of mathematics.

7.1 Pre-logicized and logicized mathematics

As a logicist, Russell claimed that mathematics is a part of logic. He claimed that mathematics, as it was taught at the time in universities and expounded in the standard textbooks, can be translated in the logical language and proved from the logical primitive propositions. The problem, which has been abundantly illustrated in what has gone before, is that one can refuse to recognize in Russell's translations an equivalent of the pre-logicized mathematical theorems. Note that the difficulty concerns the translations, not the proofs. The worry is not that Russell's proofs are gappy; it is that nothing compels us to recognize in his (well proved) theorems a correct translation of the ordinary mathematical propositions. Boutroux and Poincaré were the first to articulate this criticism. They did not contest that one can derive from the logical principles the theorems Russell claimed to have proved. What they underlined is that more is needed to show that the logical theorems express what the familiar equivalents assert. For instance, they claim that, owing to the central role the theory of differential equations plays in real analysis, an account of real analysis which, like Russell's, does not say anything about this topic is likely to be flawed. The disagreement bears here upon the definition of the mathematical content: which features of the existing mathematical concepts or the existing mathematical theories should be regarded as essential? And which features should be considered as adventitious? Boutroux and Russell had different ways of assessing what was essential to real analysis at the time. Who was right? How can we settle this sort of question?

We have reached here a murky area: it seems that one could never resolve in a satisfactory way this sort of disagreement.[1] This feeling of despondency is certainly at the source of *Faithful* and at the division of labour between mathematics and metaphysics that it induces. As we saw in the last chapter, at least since Benacerraf, one standardly distinguishes two distinct tasks: 'the *epistemological* project of providing a foundation for [the] standard mathematical theory of [the objects in a given field of mathematical inquiry]' and 'the *metaphysical* project of explaining the nature of [those objects]' (Wright 2000). The first task (defining the relevant mathematical structures) is the responsibility of the mathematicians; the second is a matter for philosophers. Now, in this framework, the (potentially) never ending disagreements are confined to the second metaphysical stage – no such a dispute ever occurs at the mathematical level. Confining the 'murky' questions to metaphysics (and preserving in this way mathematics) was however not an option for Russell. As we have seen, the issue concerning the fit between pre-logicized and logicized mathematics was for him connected to heated discussions dividing the mathematical community. According to Russell, mathematicians themselves do enter the 'murky' area; they disagree about the way one should define the basic mathematical concepts and argue for their views. The question about the 'right' definition could not thus be sidestepped – or downplayed as a merely metaphysical one. But once again: how can we solve it?

In order to advance in our analysis, let me present two ways of settling, once and for all, the dispute. I will attribute the former view to Frege, and the second one to (the later) Wittgenstein. But historical accuracy does not matter too much at this stage. My aim is indeed to use these views as two milestones from which Russell's position could be triangulated.

Frege (1914) distinguished two sorts of definition: constructive definition, which consists in introducing a new sign in the ideography to express a sense constructed out of already given ideographic signs; and analytic definition (or logical analysis), which consists in associating 'a simple sign with a long established use' with a complex expression, made of ideographic signs, 'which in our opinion has the same sense'. Frege then remarked (1914, pp. 210–11):

Let us assume that A is the long-established sign (expression) whose sense we have attempted to analyze logically by constructing a complex expression that gives the analysis. Since we are not certain whether the analysis is successful, we are not prepared to present the complex expression as one which can be replaced by the simple sign A. If it is our intention to put forward a definition proper, we are not entitled to choose the sign A, which already has a sense, but we must choose a fresh sign B, say, which has the sense of the complex expression only in virtue of the definition. The question now is whether A and B have the same sense. But we can bypass this question altogether if we are constructing a new system from the bottom up; in that case we shall make no further use of the sign A – we shall only use B. We have introduced the sign B to take the place of the complex expression in question by an arbitrary fiat and in this way we have conferred sense on it ... If we have managed in this way to construct a system for mathematics without any need for the sign A, we can leave the matter there; there is no need at all to answer the question concerning the sense in which – whatever it may be – the sign had been used earlier ... In constructing the new system we can take no account, logically speaking, of anything in mathematics that existed prior to the new system. Everything has to be made anew from the ground up. Even anything that we may have accomplished by our analytical activities is to be regarded only as preparatory work which does not itself make any appearance in the new system itself.

For Frege then, once the logical system is completed, one can forget all about pre-logicized mathematics. The system becomes independent, and nothing in it refers to the science as it existed before. According to Frege, we do not have to justify the correctness of the logical definitions, quite simply because all the definitions are constructive (stipulative). No definition is analytic. 'If we are constructing a new system from the bottom up', we can bypass the question as to whether the pre-logicized and the logicized concepts have the same content altogether, since, in this case, the sign with the long-established use does not occur anymore. Frege had a definitive answer to Boutroux and Poincaré's criticism: the definitions are not the conclusions of an analysis of existing mathematics, but the results of stipulations. They do not have to be compared to any pre-logicized body of knowledge. This answer illustrates in a very vivid way Frege's universalist standpoint: it is because the new logical system is viewed as the science of first principles, which presupposes nothing and does not refer to any prior knowledge, that 'there is no need at all to answer the question concerning the sense in which ... the sign has been used earlier'. Of course, to convince the reader not to stop reading, Frege is compelled in the *Begriffsschrift* as in the *Grundgesetze* to refer, here and there, to some already known concepts and theories; but these didactic hints are a concession to logical purity – once the system is completed, he will throw away the ladder.[2]

One can consider Part III of Wittgenstein (1956) as the defence of another conception of the relation between pre-logicized and logicized mathematics, symmetrically opposed to Frege's (1914). In Chapter 3, I remarked that, along with Poincaré and Boutroux, Wittgenstein criticized the logicists for not being sensitive enough to the fine-grained features of the mathematics they were dealing with. Thus (1956, pp. 84, 89):

> I should like to say: mathematics is a MULITICOLOURED *mixture* of techniques of proof. (§ 46)
> If someone tries to shew that mathematics is not logic, what is he trying to shew? He is surely trying to say something like: If tables, chairs, cupboards, etc. are swathed in enough paper, certainly they will look spherical in the end. He is not trying to shew that it is impossible that, for every mathematical proof, a Russellian proof can be constructed which (somehow) 'corresponds' to it, but rather that the acceptance of such a correspondence does not lean on logic. (§ 53)

Wittgenstein did not claim that the logicist project is technically flawed; he did not even claim that there is no correspondence between pre-logicized and logicized mathematical propositions; he only said that the 'acceptance of such a correspondence does not lean on logic': there are no general criteria compelling the mathematician to recognize in the logical formulae an acceptable equivalent of the theorems he is accustomed to. In particular, Wittgenstein underlined the role played in mathematics by the notation (1956, p. 84):

> If you had a system like that of Russell and produced systems like the differential calculus out of it by means of suitable definitions, you would be producing a new bit of mathematics. Now surely one could simply say: if a man had invented calculating in the decimal notation – that would have been mathematical invention! – Even if he had already got Russell's *Principia Mathematica*... I said: whoever invented calculation in the decimal notation surely made a mathematical discovery. But could he not have made this discovery all in Russellian symbols? He would so to speak, have discovered a new *aspect*.

By stressing the fact that the discovery of the decimal notation has been an important mathematical discovery, Wittgenstein suggests that the distinction between the content expressed by a mathematical sentence and the way this content is displayed in a particular symbolism is not as innocuous as it seems to be: notational features often play, mathematically speaking, a crucial role; they cannot be changed without modifying the content expressed. Notational features should be considered an 'aspect' of the mathematical content. So understood,[3] Wittgenstein's remarks appear to push Boutroux and Poincaré's reasoning to an extreme: any modification of the mathematical 'language-games' is liable to distort the nature of what is expressed. In particular, using one and the same logical language for translating the whole of existing mathematics leads to ascribing to it a unity which it is deprived of – in this respect, to logicize mathematics is like swathing 'tables, chairs, cupboards' in enough paper so as to make them look all spherical in the end. In Wittgenstein's perspective, it is not the logician, the system-builder, who has the authority to judge whether or not the logicist rewriting is acceptable, but the working mathematicians – and, owing to the tight connection between the mathematical content and its mode of notation, it is likely that he or she will always refuse to take the logical translation as an acceptable rendering of familiar theorems and proofs.

Wittgenstein and Frege represent two opposite ways of solving the pending question concerning the way one should view the relation between pre-logicized and logicized mathematics. For the latter, once the logical system is completed, the existing mathematics should take its leave: the ordinary mathematical concepts, expressed in the usual mathematical notations, are never the yardstick by which the logical undertaking should be sized – all the logical definitions, including the definition of mathematics itself, are ultimately stipulative. For the former, the mere idea of elaborating a logical system containing the whole of mathematics is absurd, since it would by itself lead to a uniformization and then to a distortion of what one seeks to represent. Here, ordinary mathematics, as it is practised by mathematicians, is the ultimate standard – and the logician is not allowed to change anything.

In a certain sense, this is only the nth rehash of Moore's famous paradox of analysis: Wittgenstein would maintain that any informative analysis is liable to be misleading, while Frege, in order to avoid once and for all any problem, would claim that, in the end, no ideographic definition is an analysis. There is however a special feature here which makes the situation even more strained. The difficulty we are facing does not only concern the way one should balance the need to be faithful with the need to be informative, but the very question of the nature and the identity of what is analysed. If mathematics is really nothing other than logic, then it is the very idea that there is a target out there (existing mathematics) that should be, in the end, eliminated: the seemingly external target is in fact nothing other than a part of the logical system. But how can we show that such is the case without comparing the logicized mathematics to the mathematics as it is given prior to and independently of logic – and thus without recognizing the fact that there is a pre-logicized mathematics after all? I will come back to this difficulty in the conclusion. But one must keep it in mind in order to understand the real depth of Frege's solution. For Frege, if mathematics is really logic, then one should give up the idea that there is an existing mathematics to which one can compare the logical system.

Let me come back to Russell. It is clear that his position lies somewhere in between the two poles represented here by Frege and the later Wittgenstein. Russell was a logicist, and did not share Wittgenstein's belief that the project of rewriting mathematics was doomed to misrepresent its object. But he did not espouse Frege's idea according to which the logical definitions were stipulative: our previous examinations of Russell's theories of geometrical space and of quantity show that he sought to display the true content of the pre-logicized notions of space, quantity and number. Now, is there an intermediate position between Wittgenstein's radical relativism (the mathematical content adheres to the various mathematical language-games) and Frege's radical universalism (the mathematical content is entirely determined from within the logical system)?

7.2 Russell's positive universalism

Universalism in Frege and Russell's work is the subject of a still running discussion in the literature. Some commentators, following Dreben and van Heijenoort (1986, p. 44), hold that Frege and Russell's universalist view forbad them from meaningfully raising any metatheoretical question:[4]

> For Frege, and for Russell and Whitehead, logic was universal: within each explicit formulation of logic, all deductive reasoning, including all of classical analysis and much of Cantorian set theory, was to be formalized. Hence, not only was pure quantification theory never at the center of their attention, but metasystematic questions as such could not be meaningfully raised. We can give different formulations of logic, formulations that differ with respect to what logical constants are taken as primitive, or what formulas are taken as formal axioms, but we have no vantage point from which we can survey a given formalism as a whole, let alone look at logic whole ... The only way to approach the problem of what a formal system can do is to derive theorems.

Against this line, some commentators have contended that universalism was compatible with metatheory. Landini (1998, pp. 32–41), elaborating on Russell and Frege's realism, maintained that the logicists had no trouble in distinguishing logical principles with the linguistic apparatus used to represent them, and thus that metalinguistic and metatheoretical approaches were not foreign to them. Other commentators, like Blanchette (2012) and Antonelli and May (2000), distinguished different versions of universalism to show that only an extreme interpretation (which it would be unreasonable to attribute to Frege or Russell) is incompatible with a metatheoretical perspective. Here however, I am not interested in the issue as to whether or not Russell and Frege could have raised standard metatheoretical questions concerning the soundness, the completeness, the independence, etc., of a formal system. My aim is different: I would like to examine how Russell and Frege's universalism impinges on the way they conceive the relations between pre-logicized and logicized mathematics.

More specifically, I will argue that a distinction, first made by Rivenc (1993), between two distinct kinds of universalism, a 'negative universalism', represented by Frege and the early Wittgenstein, and a 'positive universalism', espoused by Russell and Carnap, helps us to clear the ground. Rivenc characterized negative universalism in this way (1993, p. 26):[5]

> The decisive gesture [of negative universalism] is to label as necessarily absurd, because transgressing the laws of sense which are imposed by the nature of its basic elements, any discourse that would talk about these elements ... or more generally any discourse which attempts to enumerate these laws.

And he described positive universalism in this way (1993, p. 27):

> But logical universalism can take another form ... which I call, for want of a better term, positive universalism. Two traits may provisionally characterize it. First, the tendency to see a hypothesis that necessitates its own inexpressibility as a self-refuting hypothesis ... The other trait of positive universalism (evidently closely related to the first) is the 'finally disappearing' status that is given to preliminary explanations and clarifications. Sure, one can partly admit that there is an irreducible 'remainder' where one can make hay with the logically imperfect yet familiar ordinary language: the moment of analysis, the movement of research towards its primitive terms. But as soon as the exposition, following the order of the synthesis, is, at least ideally, in a position to extend itself ... then that which has once been expressed must be reformulated within the system, following [a manoeuvre which will be called 'internalization'].

At first, it seems difficult to spot any substantial difference between the two sorts of universalism. Rivenc seems to resume the standard (after van Heijenoort) position according to which universalists could not take logic, its 'basic elements' or its 'laws', as the subject of a discourse. Positive universalists also shared this view, since they gave to 'preliminary explanations and clarifications' (and other metatheoretical attempts to say something on logic as a whole) the status of 'finally disappearing' considerations. Thus, the sole difference between the two brands of logicism seems to be that the first asserts that any metatheoretical discourse is nonsensical, while the second holds that any such attempt appears, in the end, as nonsensical. At first, considerations about time seem completely irrelevant. But the following quote shows nevertheless that this is the role ascribed to time and progressivity which grounds Rivenc's distinction (1993, pp. 30–1):

> In contradistinction to the negative universalism of Frege and Wittgenstein, with its entrenched and permanent renunciations, positive universalism presents itself more as a movement or tendency; one could say that it is essentially motivated by the intention to internalize or that it is guided by the heuristic idea of saying everything in artificial language, in as much as it is a perfect representative of language, without judging in advance the point at which the task will (eventually) prove to be impossible, yet at the same time not feeling obliged to accomplish everything at a stroke. This trait of progressivity, if one may so speak, is one of the aspects of the general belief shared by the holders of positive universalism, that 'philosophy' is capable of progress as is science, and that the piecemeal, fragmentary and local working of science is one of the general conditions for the advancement of knowledge. The tradition of positive universalism allows provisional stages, corrections and elaborations that

are still to come; it can, then, allow that for the moment one may follow the detour of a metalanguage, that one may build a foundation on the contents already given, in as much as internalization can be only a horizon or perspective.

Russell, like Frege, was guided by the universalist ambition of saying everything in logical language, by the ambition of not referring to any external non-logical knowledge. But he did not decide, before developing the logical system, 'the point at which the task will ... prove to be impossible', nor did he feel 'obliged to accomplish everything at a stroke'. That is, Russell allowed himself to refer to some already given content, if, at a later stage in the development of the system, this content could be brought back home and constructed as an integral part of the logical machinery. Negative universalists claimed that what is nonsensical is always nonsensical, while positive universalists espoused a more liberal view, according to which one could provisionally accept the taking of an external vantage point to logic and to its principles, if further developments show that what was taken as an external point of view could be internalized in the logical system. This importance given to progressivity (one can temporarily 'open' the logical system to its outside) contrasts with the 'entrenched and permanent renunciations', which is, according to Rivenc, the hallmark of Frege's negative universalism.

I do not want, here, to defend the whole of Rivenc's analysis.[6] What I would like to stress, however, is that Rivenc's contrast sheds some light on Russell's very peculiar way of articulating pre-logicized and logicized mathematics. Indeed, in our previous chapters, we have seen that many pages of the 'remote' parts of PoM and PM (the parts where logic is confronted with 'real' mathematics) were devoted to delineating the outlines of the mathematical content. These passages correspond exactly to the 'provisional stages' that Rivenc alludes to above. In his developments concerning geometry, real analysis, the theory of quantity, the theory of real numbers, etc., Russell clearly dealt with mathematics as they were practised at the time. But for all that, these analyses should not be seen as historical descriptions of the state of the art in mathematics at the beginning of the twentieth century. As I have said, they were preparatory remarks, which were intended to support a final logical reconstruction. Thus if at first Russell's considerations appear to be external to the logical system, at a later stage all these developments are internalized in the relational framework. The distinction between positive and negative universalism provides us with the means to give a status to the part of PoM and PM that we have analysed in the previous chapters.

Let me come back to Frege. What is striking in the Frege (1914) discussion is the radical opposition which is drawn between on the one hand mathematics as it is practised outside and prior to the logical system and on the other hand mathematics as it is done within the logical system. Once the logicist task is completed, once the whole of arithmetic is logicized, said Frege, all the questions concerning the adjustment between *analysans* and *analysandum* entirely disappears. Boutroux's Wittgensteinian objection, consisting in pointing out that there is more in ordinary mathematics than in its logicized counterpart, could be neglected, since the old mathematical symbols are no longer used in the new logicized mathematics. Frege adopts then a genuine all-or-nothing strategy: as one can never be sure that the logical representation captures all the content of the mathematical target, one should not attempt to answer the question of its adequacy until the system is completed – and once the logical construction is finished, one no longer needs to meet this issue. Frege's position is thus a paradigmatic illustration of what Rivenc called negative universalism: Frege does not allow 'provisional stages, corrections and elaborations that are still to come', and this sets Russell apart from Frege, since, as we have just seen, Russell justified his logical definition by a preparatory discussion of existing mathematics.

But Russell's position must not yet be confounded with Wittgenstein's standpoint. Russell is a universalist: the importance conferred to the pre-logicized mathematical knowledge is only temporary. This means that, for Russell, the decision to count some features of the existing mathematics as an integral part of the mathematical content should be ultimately grounded on a logical analysis. As Rivenc wrote, 'as soon as the [systematic] exposition is at least ideally, in a position to extend itself ... then that which has once been expressed must be reformulated within the system'. The need for the logical definition (called 'internalization' by Rivenc) is thus what differentiates Russell from Wittgenstein. For Russell, it is not true that definitions should not change anything to pre-logicized mathematics. As we have seen, logical analysis is not a mere recording device, whose only task is to capture faithfully what is already there. Of course, the difficult problem is to determine, in each case, the extent to which transformations due to logicization are acceptable. And Wittgenstein is right to note that there is no general rule to answer this question. But it is one thing to say that there are no general criteria to settle the issue and another to say that there is never any reason to accept, in a given case, a particular rephrasing.

To summarize then, following Rivenc's suggestion, one can present the relations between Wittgenstein, Frege and Russell's positions in the following way. The Fregean considers that the perspective one gets *ex post*, once the logical construction is completed, is the only valuable one. On the contrary, the Wittgensteinian claims that the only important standpoint is the one one has *ex ante*, before the process of logicization takes place. For Russell, neither of them is right: the two perspectives, *ex ante* and *ex post*, are legitimate. What appears, at one stage of the system, as something which is given from outside (for instance, the fact that geometry should be distinguished from analytic geometry, or the fact that real numbers should be connected to measurement), becomes, at a later stage, a part of the logical machinery (these 'facts' are reflected in the respective logical definitions of these concepts). One can of course adopt the *ex post* view and consider Russell's definitions as mere stipulations which will be universally accepted once the system is completed. Or one can espouse the *ex ante* view and read Russell's construction as a historically informed examination of the mathematical practices of his time. But to really understand what Russell does, one must move from one perspective to another. Good readers of Russell should then have a sense of anticipation and a good memory. They must understand that a detailed discussion about some given mathematical data has to be seen as a preparation to a yet-to-come logical definition (it is not a mere description of some interesting aspects of a particular mathematical practice); and conversely, they must understand that the logical characterizations, far from being the results of a stipulation, proceed from an analysis of the mathematics of their time.[7] One should then take seriously the positive universalist idea that time and progressivity matter, the idea that what is yet to come is neither nothing nor already done. Indeed, this insight allows us to make room for a middle term between Wittgenstein and Frege and to understand why one should neither sacrifice the *ex ante* (Wittgenstein's) to the *ex post* (Frege's) perspectives, nor reject logicism altogether.[8]

One could say the same thing otherwise by deepening the difference between Russell's (on the one hand) and Frege and Wittgenstein's (on the other) view of the relationship between the logical system and its outside. As I explained above, there is a sharp and enduring opposition in Frege (1914) between pre-logicized and logicized mathematics. The Fregean strategy is to isolate the logical system and to maintain that, once completed, the system would supplant ordinary arithmetic. If Wittgenstein, on this last point, takes the opposite stand (ordinary arithmetic is used as a yardstick to assess the logical construction), he nevertheless shares with Frege the idea that the logical system is opposed to existing mathematics. Strange as it may seem, there is no such contrast in Russell: the boundary between the system and its outside goes inside the system. The development of the system is not viewed as an autonomous growth, unconnected to its environs, but as the result of the progressive integration of what was considered at an early stage as something external. The boundary between what is internal and what is external is then not fixed once and for all; it evolves as the system shapes up. Let's take some examples. The remarks on the distinction between pure synthetic and Cartesian geometry in PoM seems at first to be a historical consideration about different ways of doing geometry; but once the definition of projective space is adopted, one realizes that Russell's contrast is grounded on a distinction between relational types (between order and incidence relation) and that they should thus be granted a logical status. At the beginning of PM VI, the claim that real and rational numbers should be linked to measurement is grounded on a non-logical desideratum (*Applic*); but as soon as the relational definitions of numbers and vector-families are adopted, the idea that numbers (as relations of relations) are made for the measurement of quantities (which are relations) must be considered as a logical fact. In each case, what appears *ex ante* as an external datum is, *ex post*, integrated into the logical framework.

As Blanchette (2012, ch. 7) has noted, 'the spatial metaphor, according to which there is no "vantage point", no "external standpoint", no "perspective" from which to carry out metatheoretic reasoning is pervasive' in the discussion revolving around universalism. I would like here to seize this opportunity to propose a variant to the standard geographical analogies. Some historians (supporters of the new 'world history') have recently contrasted two models of the state: the nation-states, 'based on the idea of a single people in a single territory constituting itself as a unique political community', and the empire-states, in which 'different peoples within the polity [are] governed differently' (Burbank and Cooper 2010, p. 8). The relation between a nation to its environs is completely different from the relation between an empire and its margins: '[a] nation-state tends to homogenize those inside its borders and exclude those who do not belong, while the empire reaches outward and draws, usually coercively, peoples whose difference is made explicit under its rule' (ibid.).[9] One could draw an analogy between these two kinds of sovereignty and the two brands of universalism I have just contrasted. For Frege like Russell, logic rules mathematics as a state rules its people. But Frege espoused the nation-state model, while Russell was closer to the empire state paradigm.

For Frege, the differences between existing branches of mathematics were not taken into account in the logical framework, and the system, as a homogeneous whole, was constantly opposed to its environs. On the contrary, Russell self-consciously maintained the diversity of the various branches of mathematics he incorporated into his system, and the boundary between the logical framework and its outside was never fixed once and for all. As I have explained, the growth of the system, like the growth of an empire, corresponded to the progressive integration of pre-existing territories. Comparisons are odious of course. And the logical and geopolitical contexts are too different to pursue the parallel any further. But this reference to the distinction between nation and empire could nevertheless help the reader to understand the importance of what is at stake in the distinction between negative and positive universalism. Paradoxically, it could also help him or her to keep a distance from the spatial metaphor that the debate on universalism is replete with – a good way to get rid of similes is indeed to complicate them until they reach their limit.

7.3   Logicism and the definition of the mathematical content

In this section, I would like to relate what we have just said about Russell's positive universalism to his conception of the mathematical content. In Chapter 3, I showed that, in PoM, Russell, far from making mathematics uniform, ascribed to each independent branch of the subject (or more precisely, to what he considered as an independent branch) a particular relational type. I insisted on the fact that this concept was fundamentally hybrid: a relational type is a logical notion, but it also refers to a certain state of development of mathematics. There are indeed some relational forms which do not correspond to any historically constituted mathematical field (like the connected relations); conversely, certain mathematical domains are not correlated to any relational type (like the theory of differential equations). Thus to delineate the various relational types which composed the mathematical sciences, Russell had to take into consideration both the logical framework and the actual state of mathematics. This holds for PM as well. As we have seen in Chapter 6, the distinction between real analysis and the theory of real numbers was grounded both on a distinction between two kinds of relations (order relations and the relation of relations that numbers are) and on an analysis of the reception of Cantor's works.

The idea that relational type is a hybrid concept sets Russell apart from both Frege and Wittgenstein. With Frege, Russell thought that only logical differences (only differences between relational types) should be taken into account in the logical system. But with Wittgenstein, he agreed on the fact that the various mathematical topics are in some sense given in the pre-logicized mathematics. The theory of relations was used by Russell as a means to represent fine-grained differences in the mathematical practices, but it would nevertheless be a mistake to see logic as a mere receptacle of distinctions already given in existing mathematics. Logic played an active role in the delineation of the mathematical content. The Russellian way should thus be seen, once again, as a middle course between Wittgenstein's relativism and Frege's negative universalism. But is there room for such an intermediate position? To the question, how to define the real structure of mathematics, Russell answered: by looking at existing mathematics and logic. But what conceptual weights are we supposed to attribute to each of these factors? In so far as he took into account the given specificities of the mathematics he was analysing, Russell was close to the Wittgensteinian relativist position. But the role he gave to the theory of relations in the shaping of the mathematical data sent him back to the Fregean universalist position. Russell appeared then to oscillate between the two approaches, without ever reaching a stable position. For him, the logical analysis should be sensitive to the specificities of the topics considered; but the notion of a topic was ultimately fixed through the logical analytic process. But how can we avoid the circularity here? How can one justify a logical analysis by referring to some mathematical facts, which are themselves selected just because they fit in the logical setting?

There is indeed a circle in Russell. But what this circularity shows is only that one must abandon the idea that there are some general criteria enabling us to determine mechanically the identity of the different mathematical topics. The identification of the subject matter of a given piece of mathematics, the selection of the relational forms which constitute the fundamental relational types, are not the results of an application of some already predefined criteria. It is the result of a *reflective equilibrium* – a matter of the mutual support of many heterogeneous considerations.[10] If the pre-logicized mathematics is the gauge that measures the relevancy of its logical counterpart, the logical theory of relations is the framework which enables Russell to bring together different sorts of considerations and integrate them into one coherent whole. The adjustments needed to reach this equilibrium cannot be anticipated – they depend on the material available and also on the places we kept for them in the building. Sometimes, the material should be worked again to meet the logical demand; sometimes the shape of the system should be altered to make room for the recalcitrant. In any case, the equilibrium reached always proceeds from a circular back-and-forth process of adjustment between logic and existing mathematics. It is then always unstable – always liable to be upset by further examinations of the mathematical data and/or the logical resources.

The idea that the determination of the logical content is not the application of general criteria, but the result of a circular back-and-forth process, mixing logical and mathematical considerations, is something that both Wittgenstein and Frege rejected. For Wittgenstein, those who are in the position to determine whether a given translation captures mathematical content are the mathematicians; for Frege, on the contrary, the logicians always have the final say on the definition of the mathematical content. Despite their oppositions, both philosophers refuse to enter the Russellian circle. They both consider that there is no need to oppose existing mathematics to logic. Why not espouse their views? Why attempt to hybridize the mathematical discourse with the logical one?

To say, with the Wittgensteinian, that the working mathematicians always have the authority to assess the acceptability of the logical rendering presupposes that mathematicians always agree on the way one should define mathematical content. As we have said, this is not the case: there is often no consensus in the mathematical community on the question as to whether a given distinction should be regarded as mathematically relevant or not (think of the various definitions of projective space or of real numbers). What is to be faithful to the mathematical practices when mathematicians disagree with each other? Wittgensteinians, like the structuralists and neo-Fregeans who adhere to *Faithful*, idealize the harmony one encounters in the mathematical community. And this idealization is liable to lead them to assume uncritically and dogmatically that only one structuring of the mathematical field is correct, without even giving voice to the many possible alternatives. Fregeans are not on a better footing. They ground their refusal to compare pre-logicized and logicized mathematics on the idea that such a confrontation will inevitably blur the purity of logical science. But, as a matter of fact, in Frege as in Russell, logic did not by itself determine the lines of its development; the logical construction was towed along by the consideration of existing mathematics (arithmetic, in the case of Frege). Frege did not deny this, but, in Frege (1914), he downplayed the importance of these sideways looks at mathematics that he had had to take. In so doing, he denied himself any chance to consider and discuss the different ways of organizing the mathematical material and of developing the logical system. The danger is once again the same: the uncritical espousing of one possible road without even considering the other available paths.[11]

According to Russell, mathematics is a science 'where one never knows what one is talking about' (1901b, p. 266). There are many ways of delineating the mathematical content; there are many ways of carving out the mathematical field. Wittgenstein's and Frege's common mistake was to believe that mathematics is conceptually transparent, that mathematicians always know what they are talking about – or in other words, that there is never the need to discuss how one should define the mathematical concepts. This a serious mistake, since it leads us to dogmatism, that is to espouse uncritically one view without even recognizing the possibility of endorsing another. What is distinctive of Russell's programme is the idea that logic supplies a common framework in which the different possible definitions of the mathematical content could be compared and discussed. As I have shown in the previous chapters, Russell, in PoM as in PM, compared different characterizations of the basic mathematical concepts. The theory of relations enabled him to articulate different points of view and to look for a reflective equilibrium between different positions. Of course, one never knows in advance that one will finally come to a considered agreement. But one does not know in advance either that one will not. Most of all, one does not have the choice: neither Wittgenstein's nor Frege's view is an option, since there is not a unique way to construe the mathematical content, and so one cannot escape discussing the various alternatives.

There is an irreducible circle in Russell's reasoning. His analyses are justified by reference to some mathematical facts, which are brought forward because they fit in the relational framework. The whole process could then be exposed as a sleight of hand. And it is true that it is always open to becoming a mere trick. But the important point is that it is not doomed to be one: the success of the project entirely depends on the nature of the equilibrium obtained and on the way it is defended. In other words, the circularity in Russell's reasoning is nothing else than an abstract description of the arguments expounded in PoM and PM's 'remote' parts, and which has been examined in the previous chapters. To assess their strength, one must reinsert these arguments in their concrete contexts and take into account the particular features of Russell's analyses. Thus, to form a considered judgement and to probe our intuitions, one has no other option than to redo for oneself the complicated journey that Russell invited us to undertake. One cannot short cut the intricate and dry explanations Russell gave us in PoM VI and PM VI. The solution of the circularity problem was already in our hands. What misled us was our wish to find some 'general criteria' allowing us to assess the results of an analysis. There are no such things. There is no recipe to combine the *ex ante* (mathematical) and *ex post* (logical) perspectives. And in a sense, this is as expected. Analysis is a difficult task, and it would be very strange if a uniform mechanical process could determine in a univocal way, independently of any consideration relating to the topic, how one should analyse mathematical concepts.

To give some flesh to what we have just said, let me then come back to our two main examples: Russell's analysis of projective geometry in PoM VI and his theory of real numbers and quantity in PM VI. Russell did not consider his definition of projective space to be an arbitrary stipulation. As I have indicated, his characterization leant on an analysis of the geometry of his time. In particular, he endorsed Von Staudt's view about the independence of the projective method and was very influenced by Klein's projective derivation of the standard non-Euclidean metrics. Thus, the central place he granted to projective geometry and his emphasis on incidence did not come out of the blue. At the same time, it is clear that the development of geometry did not compel him to espouse this definition of projective space. As Russell made clear, other options were available – Pasch's view based on order, the so-called 'Leibnizian' approach, etc. Russell's choice was not completely determined by the practices of the working geometers – some other constraints, coming from his relational framework, were taken into account by him. Thus, Pasch's ordinal foundation brought geometry much too close to real analysis, and the relation of distance was too complicated to be regarded as a relational type. Russell's characterization of projective space was not an arbitrary stipulation (as the Fregean would have it); neither was it a mere recording of what the working mathematicians did (as the Wittgensteinian would say). Russell attempted to develop a fragile equilibrium in which the considerations coming from different sources fitted together in one coherent whole. Of course, the delicate balance he put between the various arguments could be challenged. One could consider for instance that the importance he gave to projective geometry was excessive, or that the picture he drew of this theory did not attach enough weight to complex projective geometry. One could favour a more simple logicism, where geometry would be developed according to the Cartesian method. There are no definitive ways to meet these criticisms. But to acknowledge this does not mean that the reasons Russell gave in favour of his conclusion have no value. The way his analysis combines very different threads (the theory of relations, Von Staudt's programme, Pieri's work, Whitehead's theory of extension, Klein's derivation of metric, Poincaré's argument against the idea that rigidity is a mechanical concept, the view that metrical geometry is empirical) in one coherent whole remains impressive.

The same holds for the theory of real numbers in PM VI. Russell and Whitehead's theory was an attempt to bring together four different developments: the idea that rational and real numbers are tightly connected to measurement of quantity, the claim that the arithmetical properties of rational and real numbers must be grounded on logic, the relational theory of magnitude, and the view that real analysis has nothing to do with real numbers (with real valued functions). Russell and Whitehead succeeded in giving room to these different requirements, and what is beautiful in the construction is the way the logical resource is used to attain this goal. Logic did not act as a steamroller, which smoothed away every specificity; on the contrary, it was used here to integrate seemingly contradictory demands (the claims that real numbers are logical objects, that they should be connected to measurement, that quantities are relations, that numbers are dispensable in real analysis) into one coherent whole. Now, once again, despite its strength, one could remain unconvinced by PM's achievement. One could, for instance, argue that the PM theory does not take sufficient account of the central importance of the theory of real-valued functions, or that it is much too complicated, or that logicism does not need to account for real numbers or for its application. But that the theory could be legitimately criticized does not show that it is worthless; it only means that it is the result of a complicated back-and-forth process, a balance of considerations that one could always choose to weigh up in a different manner. Above all, what is important is less the result than the intent: that is, to use logic as a framework for discussing and arbitrating between different ways of carving out the mathematical content.[12]

7.4   Russell's logicism and practice-based philosophy of mathematics

Until now, my investigation has been mainly historical: I have examined certain neglected parts of PoM and PM, with the prospect of revising the received view of Russell's logicism. But are there any more general philosophical lessons to extract from our previous discussions?

At first sight, this does not seem to be the case. Indeed, I have portrayed Russell as a sort of craftsman, whose art essentially depended on the tools and materials he had at his disposal. I have shown that his logicism was arranged to meet certain constraints imposed by the contemporary state of mathematics and logic, and I have claimed that the conceptual strength of his arguments draw their greatness from this adjustment. It would then be a serious mistake to believe that his results could be transplanted to our present-day situation. Indeed, since Russell's time, both the mathematical world and the logical toolkit have considerably changed. Nobody today would put projective geometry at the centre of the geometrical building, and the present-day theory of measurement has nothing to do with what is presented in PM VI.[13] More importantly, the very attempt to give an overview of the whole of mathematics seems now to be too demanding: the subject has fragmented so much during the last century that nobody seems today to be in a position to give even a rough survey of its arborescence. Logic, for its part, has also considerably evolved. The distinction between syntax and semantics, the rise of the model theory, Gödel's incompleteness theorems, etc., can obviously not be ignored – and the relationship between Russell's logic and ours poses a host of difficult problems.[14] Owing to this, it seems that any attempt to extend Russell's conclusions to our present day situation is a sheer nonsense. In any case, my reading of Russell does not supply this kind of goods.

If a lesson is to be extracted from what I have said so far, it would be essentially negative. Russell's philosophy of mathematics is tightly connected to a state of mathematics which no longer exists and one cannot draw any straightforward conclusion on today's mathematics from his works. But the way he used logic as a framework to discuss the mathematics of his time seems to me still valuable. In particular, his positive universalism could shed a different light on the growing opposition between the so-called practice-based and foundational approaches in philosophy of mathematics.

In the last two decades, more and more philosophers have expressed dissatisfaction with what they experience as an estrangement between philosophy of mathematics and the practices of working mathematicians.[15] Their basic claim is that the exclusive concern for foundational issues led mainstream philosophers to focus on some topics, which, despite their importance, are not representative of what mathematicians usually do. Foundationalists focus their attention exclusively on certain mathematical theories, like arithmetic or set theory, whose truth is said to imply the truth of the other mathematical fields. The 'foundationalist filter'[16] therefore splits mathematics into two distinct parts: on the one hand arithmetic and set theory that is worthy of philosophical consideration, and on the other hand the rest of mathematics (algebraic geometry, non-commutative algebra, the theory of partial differential equations, etc.) which has no foundational importance. This amounts to clearing many potentially important topics off the philosophical table. Thus, Mancosu (2008, p. 1) explains how the foundational quest (he speaks more precisely about Benacerraf) has diverted philosophical eyes from the more advanced parts of mathematics:

The agenda set by Benacerraf's writings for philosophy of mathematics was that of explaining how, if there are abstract objects, we could have access to them. [A consequence] of the way in which the discussion has been framed is that no particular attention to mathematical practice seemed to be required by an epistemologist of mathematics. After all, the issue of abstract objects confronts us already at the most elementary levels of arithmetic, geometry and set theory. It would seem that paying attention to other branches of mathematics is irrelevant for solving the key problems of the discipline. This engendered an extremely narrow view of mathematical epistemology within mainstream philosophy of mathematics.

The upholders of the practice-based approach, in a revolutionary move, argue that if the consequence of the foundational programme is to lose contact with the real mathematics, then one should abandon it. As Corfield writes (2003, p. 269): 'one can say with very little fear that in today's philosophy of mathematics, it is the philosophy which dictates the agenda. The issue at stake ... is at what point does it becomes incumbent upon philosophers to take the reverse attitude and let the mathematics have some say in what is asked of it'.

At first, it seems that what I have presented in this book is all grist to the practice-based philosopher's mill. Russell is often depicted as a paragon of foundationalism, and targeted as such in the practice-based literature.[17] Now, what I have shown in the previous chapters is that he tried to capture as much of mathematical practices as he could. He based his logicism on a detailed analysis and reconstruction of significant parts of the mathematics of his time. Of course, one could criticize some of his conclusions – but it would be unfair not to recognize that he attempted to take into account some fine-grained features of the mathematics he was confronted with. If the picture of a Russell only concerned with logical issues and indifferent (if not ignorant) to real mathematics has been prevalent, it is first and foremost because PoM and PM have not been read to their ends. The reassessment of Russell's works about geometry and theory of magnitudes (notably the reassessment of the anti-arithmetization components of his thought) makes us realize that the depiction of Russell as a narrow foundationalist is nothing but a caricature. He did not share the view according to which only arithmetic and set theory (let's say) are worthy of philosophical consideration. On the contrary, the real Russell emphasized that the logical reconstruction should account for the independence of geometry and for the connection between real numbers and measurement. One could even say that the distinctive kind of logicism defended by Russell anticipated the recent practice-based approaches.

But a second look shows that the situation is slightly more complicated and interesting. We have seen that the practice-based philosophers seek to reverse the philosophical agenda and to 'let the mathematics have some say in what is asked of it'. In this respect, the practice-based approach is close to the view espoused by the foundationalists Shapiro, Hale and Wright, and encapsulated in *Faithful*: the philosophers of mathematics should take in the information given by the mathematicians without attempting to change it. But is it so easy to fulfil this task? Where are we supposed to find what mathematics has to say about what is asked of it? The foundationalist answer, the one displayed in *Faithful*, is that all the relevant information is contained in the syntax of the mathematical discourse. Practice-based philosophers disagree – the whole point of focusing on practices is precisely to show that many important features of the mathematical activities escape formalization. But practice-based philosophers do not clearly specify where we should look to get a hold on these 'practices'. Should we focus on the mathematical textbooks? Well, I'm not sure: textbooks give no information about the ways mathematical knowledge is extended. They provide us with the end-product, not with the complicated process which leads to them. Should we then turn our attention to the mathematical research papers? But which fields of research are we supposed to choose? As I said, mathematics is today in a very scattered state. So how can we justify the study of one topic to the detriment of others? Moreover, mathematicians often stress the fact that the more interesting part of their activity is hidden in the final published papers, so that, even if the question of the choice of the topics could be satisfactorily solved, the focus on publication would make us miss some important aspects of mathematical practices. Following Hadamard (1945), we could attempt to concentrate on the psychology of mathematical discovery – or on the cognitive setting of the use of certain mathematical notions.[18] Or we could try to base our understanding of mathematical practices on a sociological investigation of the communication networks (the study of mathematical correspondences, informal discussions, emails, etc.).[19] All these paths have been and still are followed. My argument is not that they are dead ends, but that the mere call for a return to mathematical practices is in itself too vague and ambiguous to provide us with any positive insight.[20]

One could reach the same conclusion by a different path. The methodology of case studies plays a very important role in the practice-based philosophy of mathematics. Most of the time, practice-based philosophers ground their investigation on a particular 'case' – on a particular topic (e.g. mathematical explanation), or on a particular historical sequence (e.g. the rise of conceptual mathematics in the works of Riemann and Dedekind), or on the use of a particular technique (e.g. computer-assisted proof) – and draw some general conclusions from its detailed study. This method, in which one starts from the examination of a piece of mathematics, seems to satisfy the demand not to let philosophy dictate the agenda. But it poses at least two difficulties. First, upholders of the case-study methodology assume that the piece of mathematical practice (the 'case') examined can, without damage, be isolated and detached from its environs. Second, they presuppose that one can draw from the case at stake some conclusions which hold for mathematics in general. These two assumptions are open to discussion. Against the first one, one could point out that, often, a certain mathematical notion, looked at from a certain context, appears different when viewed from another angle. For instance, real numbers, when related to the introduction of coordinates, do not look exactly the same as when connected to real analysis. Now, the question of determining which is the best framework to articulate a given concept is usually not settled by the mathematicians. The way one delimits the case one wants to study is thus not neutral: one can neither divide at will the mathematical material, nor rely on a consensual pre-definite division. But even if one assumes that one could pass over this difficulty, the second assumption poses new problems. Practice-based philosophers are not historians: they usually claim that the conclusions reached in a particular case can be extended beyond the case considered. But they usually do not argue for that – that is, they do not specify the place this case fills within the global mathematical setting. How are we then supposed to avoid the danger that the philosophy of mathematics becomes an endless collection of scattered local studies, which does not provide us with any overview of the multiple and fragmented perspectives on mathematics that we are presented with? Worse, how can we rule out the temptation to generalize in one stroke the conclusions extracted from a very particular case?

My point is not to say that the methodology of the case study is worthless. On the contrary, in previous chapters I have stressed the importance of taking into account the studies of existing mathematics in logical reconstruction. My point is to draw attention to the fact that, at some stages, case studies must be integrated into a more general framework. The need for an articulation between local studies is sometimes recognized by the practice-based philosophers.[21] But they seem to view this task as something which should be done at a second stage, only once enough material has been piled up. This I find unconvincing. As we have seen, the very decision to study one particular topic, one particular period, one particular technique, is not philosophically neutral – it commits us to organizing mathematics in a certain way, and straightaway puts some constraints on the way the different cases should be related. There is no philosophically neutral way of dividing mathematics in the cases which we can rely on. Mathematicians never stop arguing about this issue, and the ways we carve out our cases already commits us to favouring certain alternatives over others. One cannot first accumulate the material and then think about its structuring. The view that, by studying 'cases', mathematics could be caught in its pure and natural state, untouched by any philosophical distortion, is then nothing but a delusion.

This criticism of the case-study methodology is of course in line with my reading of Russell. In a certain sense, contemporary practice-based philosophers share with Wittgenstein the view that the way the different mathematical 'subjects' (be they called 'mathematical practices' or 'mathematical language-games') are particularized is immediately given in the existing mathematics. As we know, Russell did not share this optimism. He did not think that the task of the philosopher is first to recollect the evidences which are implicit in the mathematical practices, and then, in a second moment, to integrate the various data into one coherent whole. On the contrary, for him, the two tasks had to be performed at the same time – the mathematical data were to be adjusted to the logical scaffolding, which, in its turn, should be developed so as to be capable of integrating the data. Thus, for instance, in the course of his study of projective geometry, Russell justified the decision to devote an entire part of PoM to the investigation of projective geometry, by correlating projective space to a specific relational type. In Russell, the detailed study of a mathematical case was always combined with an architectonic reflection on the relations between the case and its surroundings. It is at this stage that the logical framework played a decisive role. As I explained in the last section, the logical setting allowed Russell to introduce a confrontation between the different possible structurings of the mathematical material – to compare and balance the different views that mathematicians had about mathematics. What is missing in the practice-based approach is precisely a common framework within which the various possible ways of dividing mathematics into 'cases' could be discussed.

I have said that the lessons one could extract from my interpretation of Russell's works are mainly negative. But negative conclusions, words of caution, are not always useless. The diagnosis made by the practice-based philosophers is right: the growing estrangement of mathematics and philosophy of mathematics is certainly the symptom of a deep crisis within philosophy of mathematics. But my reading of Russell leads me to think that the plea for a return to real mathematics, far from being an appropriate medicine, is only another symptom of the problem. Wittgensteinians are prone to denounce the tendency of the philosophers, and especially of the foundationalist philosophers, to mythologize certain distinctions and definitions.[22] But the concept of 'real mathematics' or 'mathematical practices', which one is supposed to come back to, is itself a piece of philosophical mythology. Mathematics is full of foundational disputes over the correct definition of a concept, over the right proof of a theorem – and the desire to come back to a mathematical state free of any philosophy is thus nothing else than the product of the philosophical fantasy, divorced from the realities of the mathematical world. In this respect, the practice-based philosophers resemble city dwellers, who fantasize about returning to the countryside in order to find the 'real' nature, the 'real' life, the 'real' animals, without realizing that this idealized countryside exists only in their imagination and is nothing else than a product of their urban life. And one fears that the call for a return to real mathematics produces the same effects as the call for a return to nature: disillusion.

To resume Corfield's terminology, one cannot do, in philosophy, without a 'filter'. Not every feature of a given mathematical practice should and could be taken into account in the philosophy of mathematics. The real issue is thus not whether one should use a filter; it is to find an agreement about its calibration. The tuning should neither be too thin (since then too many topics will be cleared off the philosophical table) nor too coarse (since then the indefinite multiplicity of cases will forbid any articulation between them). In Russell, the logical scaffolding supplied a common setting within which one could discuss this issue – within which one could experiment and probe different ways of calibrating the philosophical filter. This setting should not necessarily be of a logical nature. In the history of philosophy, there may have been some other kinds of conceptual framework within which a fruitful interaction between mathematics and philosophy developed.[23] Moreover, it is likely that Russell's logic no longer provides us with the right setting for a philosophy of today's mathematics. What I am saying, though, is that one cannot do, in philosophy of mathematics, without a common setting that allows us to deal with the calibration issue. It seems that, today, no such setting is available. Practice-based philosophers are right: the toolkit used to frame the philosophical agenda does no longer bite into the mathematical meat. But if this is true, one would do better to strive to develop a more flexible and articulated philosophical framework than to chase after the elusive 'real' mathematics.

7.5   Conclusion

I have made in this book a more extensive use of the history of mathematics than usual in the commentaries of PoM and PM. It is important not to misunderstand this use. In what precedes, I have not referred to history to explain Russell's opinions, but to identify the issues he was confronted with. More precisely, historical contextualization has enabled me to show that, at the time of PoM and PM, mathematicians did not agree on the way they should view the architecture of mathematics. Arithmetization was not accepted by everyone; geometry was construed in a lot of different ways. Russell had then different options for developing his logical system, as various definitions and several structuring schemes were on hand. His problem was not only to find a path from logic to mathematics, but also to choose between different ways of constructing the mathematical sciences. My main claim is that Russell used logic as a frame for discussing the various ways of expanding logic as a tool for reaching a reflective agreement on the way mathematics should be defined.

This sets my reading of logicism apart from the standard one. Indeed, logicism is usually presented as the claim that one can prove all the mathematical propositions from logical principles, without resorting to any kind of intuition. Logicism is thus viewed as a technical result (a theorem), which can be either true or false,[24] and whose very formulation presupposes that a definite content has been given to the mathematics one is supposed to reach. I have insisted here on another aspect: the mathematical sciences that the logicist speaks about have to be defined; and this definition is an integral part of the logicist undertaking. It would then be a mistake to believe that the goal of the logicist venture is already set, and that the only issue is to know whether logic is strong enough to allow us to reach it. On the contrary, the logicist must both prove that mathematics is a part of logic and make its meaning definite. In the logicist journey, the destination is not given beforehand; it is discussed on the way. In the remote parts of PoM and PM, logic was not used to derive the main mathematical theorems. It was used to discuss and appraise the various ways of organizing and carving out mathematics. Of course, as I have said, Russell's conclusions were open to debate: one cannot prove that a given definition is the right one, and that the destination reached is the one we had to head for. But this does not mean that one cannot come to a reflective agreement on this question, that one cannot bring together and balance the many different insights which conflict in existing mathematics.

Russell's demand for justifying the definitions of mathematical concepts, his wish to embrace the mathematical practices and the issues which divided the mathematical community, also distinguished his logicism from Frege's. Taking up Rivenc's insightful distinction, I have insisted on the fact that, unlike Frege, Russell did not oppose the logical system to the pre-logicized mathematics. In PoM and PM, the distinction between what is outside (the existing mathematics) and what is inside (logic) the system is a moving demarcation line, whose outline evolves as the logical framework shapes up. What appears at first as an external content is considered as a part of logic at a later stage; and conversely, the logical definitions are always rooted in preliminary discussions. The remote parts of PoM and PM, which we have examined in detail, represent the 'wild frontier', where Russell confronted the logical scaffolding with the mathematics of the working mathematicians, with the prospect of reaching a considered agreement. These developments have no equivalent in Frege – or at least, Frege (1914) firmly refused any attempt to compare pre-logicized and logicized mathematics.[25]

I hope that the reader, by the end of this book, better understands my strange decision to begin a study of PoM and PM by an examination of their final parts. This choice, which was justified by external circumstances (there is no commentary of Russell's theories of geometry and quantity in the literature), was also based on some substantive reasons. The distinctive features of Russell's logicism, his emphasis on progressivity and the place he ascribed to the confrontation with real mathematics, do not surface in the first parts of PoM and PM, where he developed a brand new science, the logic of relations. What one misses, when one passes over in silence the remote parts of PoM and PM, is much more than the actual doctrines expounded there – what one misses is the specific dynamic of his project; and that is what makes it still interesting today. In the Introduction, I distinguished between two kinds of analyses: the analysis of the logical notions (expounded in the first parts of PoM and PM) and the analysis of the mathematical concepts (elaborated in the remote parts). There seems to be at first no reason to make such a distinction, since, according to Russell, mathematical concepts are, ultimately, nothing but logical notions. But for him, the ultimate perspective, the one one gets at the end, is never the only one to consider. The perspectives one obtains on the way also have their significance. My opening distinction between logical and mathematical notions was based on nothing else than time considerations. But time mattered for Russell.

Notes

Introduction

1. On Russell's notion of analysis, see, among others, Hylton (1990), Griffin (1991), Hager (1994), Landini (1998; 2007), Linsky (1999) and Levine (2009).
2. See Hylton (1990), Griffin (1991) and Levine (2009).
3. See Hylton (1990), Landini (1998) and Makin (2000).
4. See Griffin (1985) and Stevens (2005).
5. See Landini (1998) and Linski (1999).
6. See Hylton (1990), Landini (1998) and Linsky (1999).
7. See Wahl (2007) and Levine (2009).
8. There are notable exceptions. Grattan-Guinness (2000) devotes a few pages to both Russell's theory of geometry and his theory of quantity. Torretti (1978) deals at length with the theory of geometry – but he focuses his account on Russell (1897a), not on PoM. Musgrave (1977) and Coffa (1981) refer to PoM VI, but, as we will see in Chapter 2, their interpretation is burdened by some important misunderstandings. Concerning Russell's theory of quantity, the harvest is even poorer. Quine (1963) refers to PM VI, but he makes it clear that he does not aim at producing a comment that is faithful to Russell's thought and leaves aside what seems to be the most original part of Russell's and Whitehead's construction. Following Quine, Bigelow (1988) resumes some of Russell's insights, but his account of PM VI is seriously mistaken (see Chapter 4).
9. In order to emphasize the strangeness of the situation, note that, among the 200 pages that the French philosopher Louis Couturat (1905) devoted to PoM, 90 pages were dedicated to Russell's theory of space and 30 pages to the examination of *l'idée de grandeur*. More than half of the review was then devoted to topics that have completely disappeared from the subsequent accounts of Russell's logicism. Russell himself is certainly partially responsible for this neglect. In his *Introduction to Mathematical Philosophy* (1919), Russell did not speak about geometry at all and greatly oversimplified his theory of rational and real numbers.
10. See Wittgenstein (1953, §596): 'A main cause of philosophical diseases – a one-sided diet: one nourishes one's thinking with only one kind of example'.
11. The translation schema uses a coordinate system by which every point is represented by a couple of real number coordinates. Now, any equation involving the coordinates specifies a subset of the plane, namely the solution set for the equation. For instance, when a Cartesian coordinate system is used, a line in the Euclidean plane corresponds to the solution set of a linear equation involving two variables.
12. According to Benacerraf (1981, p. 46), Frege had similar concerns: '[Frege needs] an argument to show that the sentences of arithmetic, in their pre-analytic senses, mean the same (or approximately the same) as their homonyms in the logicist system'. Walsh (2010, p. 42) notes that Benacerraf took this question from Cassirer's *On the Concept of Substance and the Concept of Function*. On Frege, see pp. 190–3.
13. On universalism, see Dreben and Van Heijenoort (1986), Goldfarb (1989), Hylton (1990), Rivenc (1993), Landini (1998), Antonelli and May (2000), Blanchette (2012).

14. 'No mathematical subject has made, in recent years, greater advances than the theory of arithmetic. The movement in favour of correctness in deduction inaugurated by Weierstrass, has been brilliantly continued by Dedekind, Cantor, Frege, and Peano, and attains its final goal by means of the logic of relations' (PoM, p. 111). 'The mathematical treatment of the principles of mathematics, which is the subject of the present work, has arisen from the conjunction of two different studies, both in the main very modern. On the one hand we have the work of analysts and geometers, in the way of formulating and systematising their axioms, and the work of Cantor and others on such matters as the theory of aggregates. On the other hand, we have symbolic logic, which ... has now, thanks to Peano and his followers, acquired the technical adaptability and the logical comprehensiveness that are essential to a mathematical instrument for dealing with what have hitherto been the beginnings of mathematics' (PM, p. v).

15. This is precisely the reason why I focus here on these theories rather than on Russell's conceptions of real analysis or transfinite arithmetic. In these latter cases, Russell, in the main, followed Cantor and Dedekind – even if, as we will see in Chapter 6, this faithfulness is tempered with some important changes.

16. The surviving portion of the Russell–Whitehead correspondence remains unpublished. It is available for consultation in the Russell Archives at McMaster University, Hamilton. I thank G. Whitehead for having allowed me to use Whitehead's letters.

17. Indeed, Russell and Whitehead attempted to develop a large part of arithmetic without resorting to the infinity axiom – by using an ascent in the type hierarchy (see the 'Prefatory Statement' at the beginning of PM II, and 1919, ch. 13).

18. See for instance Boolos (1994) and Landini (1998; 2007). I will come back to this question in Chapter 3.

19. The same worry surfaced in the doctrine of (rational and real) numbers and quantity in PM VI, where Russell and Whitehead opposed to the Dedekindian definition of the real numbers. Here again, the issue was not to show that there was a way to reduce the theory of real numbers to logic, but to argue for a particular reduction against some other possible ones – see Chapters 5 and 6.

20. Benacerraf (1965) illustrated the multiple reductions problem by the different set-theoretical definitions of the finite ordinals. But neither in PM nor in PoM did Russell consider different variants to his definition of finite numbers in terms of class similarity.

1   Projective Geometry

1. There is a manuscript of more than 100 folios devoted to Klein's presentation of geometry kept at the Russell Archives at McMaster University.

2. On Russell's criticism of Kant and Lotze, see Griffin (1991). I will not comment on these chapters in this book.

3. See Pieri (1898).

4. See Pasch (1882). Russell referred as well to Peano's (1889b) version of Pasch's construction.

5. What follows is thus certainly not a presentation of the basics of projective geometry – to follow too closely the standard of the contemporary mathematics would have caused us to lose sight of the way projective geometry was done when Russell wrote PoM. For an accessible introduction to the subject, see Brannan *et al.* (1999) and Coxeter (1949). For a more detailed historical account, see Nabonnand (2008).

6. Today one views the Euclidean figures as sets of points that are invariant according to a certain kind of transformation. From this perspective, Poncelet's insight amounts to having extended the 'group' of permissible transformations. But of course, before Klein's Erlanger programme, geometers did not view space in this way.
7. One can easily show that perspectivity in the plane is defined by the transformation of two points.
8. As is well-known, the points and the line at infinity were not the only objects Poncelet was led to postulate in his study of the projective properties. He also introduced complex points, like the so-called 'cyclic points', which are the two imaginary intersections that any two Euclidean circles have in common. At the time of Poncelet, the distinction between real and complex projective space was not really made. But, by the end of the nineteenth century, the difference was acknowledged as a crucial one, and, as Russell focused on real projective geometry, I will not say much about complex projective geometry.
9. The real projective plane P^2 is usually defined today as the quotient of $\mathbb{R}^3 \setminus \{O\}$ by the equivalence relation ~, which relates two vectors **u**, **v**, when **u** = λ**v**, with λ being any real number. One finds this 'algebraic' definition as early as Veblen and Young (1910). But the use of homogeneous coordinates is common since Plücker in the middle of the nineteenth century.
10. In a projective space, two distinct coplanar lines always intersect in a point (eventually the point at infinity), and two distinct planes always intersect in a line (eventually the line at infinity).
11. PoM III is devoted to order, and in this part Russell discussed at length the notion of separation (see §§ 194, 203, 224–33).
12. A cross-ratio between two pairs of lines in a pencil can be defined as a ratio between two angular ratios. On cross-ratios (Russell used the term 'anharmonic ratios'), see PoM, §406.
13. One can see the harmonic conjugation as a generalization of the notion of bisection. To say that C is the harmonic conjugate of D is to say that if we imagine that D is the point at infinity on the line (AB), then C is the middle point between A nd B.
14. For instance, Menelaus' theorem says that if a triangle ABC is cut by a line ('transversal') in three points at P, Q and R (all distinct from A, B, C), then $PA \times QC \times RB = PB \times QA \times RC$. Now, to demonstrate that three points P, Q and R were collinear, Carnot and his followers typically used Menelaus' theorem.
15. In real projective geometry, the fundamental theorem and Pappus's theorem are interdeducible. But one can have Desargues's theorem without having Pappus's.
16. As I said, my aim here was not to survey, even succinctly, the development of projective geometry. For instance, I have said nothing about duality. Duality is of course a key topic in the development of the projective theory, but it plays only a peripheral role in PoM VI, which is why I have decided to leave it out here.
17. As we will see in the next chapter, this was the ambition of Klein (1871).
18. Pasch (1882) is regarded as the first complete axiomatization of real projective geometry. This was followed by Peano (1889b), though Von Staudt (1847) could be regarded as a first step in that direction.
19. On the relationship between Hilbert and the projective geometers, see Hallett and Majer (2004).
20. Coxeter's first seven axioms characterized the incidence structure of the projective space. As is well-known, in the plane the situation is not the same – the

planar correspondent of the incidence axioms do not suffice to prove Desargues's theorem. There are non-Arguesian planes. On this topic, see Russell's footnote to PoM, p. 309.

21. See PoM, pp. 384–5: 'the uniqueness of the above construction ... is the fundamental proposition of projective geometry'.

22. See PoM, p. 385: 'we may regard the harmonic relation, as in the symbol $cH_{ab}d$, as a relation between two points, which involves a reference to two others. In this way, regarding a, b, c, as fixed, we obtain three new points d, e, f on the line ab by the relations $cH_{ab}d$, $aH_{bc}e$, $bH_{ac}f$. Each of these may be used, with two of the previous points, to determine a fourth point, and so on. This leads to what Möbius calls a *net*, and forms the method by which Klein introduces projective coordinates'.

23. Russell grounded his development on Klein. On Von Staudt's original method of coordination, see Nabonnand (2008).

24. For a rigorous treatment, see Coxeter (1949, ch. 11). To pursue the presentation in Klein's style (which is also Russell's in PoM, §§ 368, 407), the point $1/n$ will be the conjugate of n with respect to 0 and 1. To find it, draw the line which passes through n and Q. Let N be the intersection between the line nQ and the line $0P$. Now, consider the intersection N' between the line $1N$ and the line $Q\infty$, and draw the line $N'R_0$. The intersection between this last line $N'R_0$ and the line 01 gives the point $1/n$.

25. Russell summarized the point in this way (PoM, p. 389): 'the principle adopted in assigning numbers to points (a principle which, from our standpoint, has no motive save convenience) is the following: if p, q, r be the numbers assigned to three points already constructed, and s be the number to be assigned to the harmonic conjugate ... of the q-point with respect to the p-point and the r-point, then we are to have $(p - q/r - q)/(p - s/r - s) = -1$'.

26. A subset A of a topological space X (the Möbius net) is dense in X (dense in the projective line) if for any point x in X, any neighbourhood of x (any open interval of the line) contains at least one point from A.

27. See Russell's diagnosis in PoM, p. 389: 'before proceeding further, it may be well to point out a logical error, which is very apt to be committed, and has been committed, I think, even by Klein ... Our points have no order but that which results from the net, whose construction has just been explained. Hence only rational points (i.e. such as, starting from three given points, have rational coordinates) can have an order at all. If there be any other points, there can be no sense in which these can be limits of series of rational points, nor any reason for assigning irrational coordinates to them. For a limit and the series which it limits must both belong to some one series; but in this case, the rational points form the whole of the series'.

28. Russell's use of 'descriptive' has nothing to do with Monge's '*géométrie descriptive*'. As far as I can see, the use of 'descriptive' in PoM is idiosyncratic.

29. In Pasch's initial presentation, the (limited) plane was an indefinable. For more details on Pasch's and Peano's works, see Gandon (2006b).

30. Instead of starting with the three-termed relation 'between', Russell used transitive asymmetrical relations. This does not greatly affect the content of the axiomatization – see PoM, p. 395.

31. But the notion of 'descriptive plane' (in Pasch's or Russell's sense) is more general than the affine plane, since, through any point outside a given line l, there can be more than one line which does not intersect l.

32. Pasch seems to have developed this from Klein (1873). For more on Pasch's construction and his relation to Klein (1873), see Gandon (2005b).

33. For proving the fundamental theorem (and thus Pappus's), Pasch introduced a new primitive relation: congruence between points. For reasons related to his empiricism (which he explained in 1882, §13), he did not want to set up (an equivalent of) Coxeter's continuity axiom 14. But nothing (except empiricist prejudice) forbade him from doing so, and Russell was right to say that Pasch's construction provided a sufficient basis for developing projective geometry.

34. See also p. 386: 'given any three different points a, b, c on a line, consider the class of points x such that a and c, b and x are each harmonic conjugates with respect to some pair of points y, y' ... Here y, y' are supposed variable: that is, if any such points can be found, x is to belong to the class considered. This class contains the point b, but not a or c. Let us call it the segment (abc). Let us denote the relation of b to x (a and c being fixed) by $bQ_{ac}x$. Then Q_{ac} is symmetrical, and also $bQ_{ac}x$ implies $aQ_{bx}c''$, where Q_{ac} represents the relation of non separation with respect to a and c'.

35. For more on Pieri, see Marchisotto and Smith (2007).

36. See Gandon (2004). On Russell's evolution during the period, see as well Griffin (1991).

37. For more on this story, see Carus and Fano (1915). See also Wilson's (2010) researches on the mathematical setting of Frege's thought.

38. Let me quote at length a seminal passage of Carnot (1804, p. 9): '[synthesis, i.e. the old geometric method] is restricted by the very nature of its procedure; it can never lose sight of its object: this object should always be presented to the mind in a neat and genuine way ... [Synthesis] then cannot reason about non-real quantities, non-possible operations; it can make use of some signs, which could help the imagination and memory; however these signs can be nothing but mere abbreviations. Analysis, on the contrary, has, first, all the means of synthesis, and, what is more, it allows some objects which do not exist in these combinations; it represents them by symbols, as if they were real: it mixes real being with the being of reason; then, by some methodical transformations, it manages to eliminate ... them. Then, what was unintelligible in the formulae disappears, and there only remains what a subtle synthesis would probably have discovered: but this result has been obtained by a shorter and easier, nearly purely mechanical, way.'

39. On Von Staudt, see Nabonnand (2008). Hilbert (1899, ch. 3), adapted Von Staudt's reasoning to the Euclidean framework.

40. In particular, he maintained Von Staudt's quadrilateral construction – see for instance Russell (1897a, pp. 126–8).

41. Russell's interpretation of Riemann's work is flawed. For more on this, see Chapter 2 and Torretti (1978, pp. 314–18).

42. For a criticism of Russell's position in *An Essay*, see Torretti (1978).

43. Here are the alleged 'postulates' (1897a, p. 133): 'I. We can distinguish different parts of space, but all are qualitatively similar, and are distinguished only by the immediate fact that they lie outside one another. II. Space is continuous and infinitely divisible; the result of the infinite division, the zero of extension, is called a *point*. III. Any two points determine a unique figure, called a straight line, any three in general determine a unique figure, the plane.'

44. Poincaré (1902, p. 60) developed this idea.

45. This manuscript has been published in volume 2 of *The Collected Papers of Bertrand Russell*.

46. The first axiom explains how to connect the points to define a line, to connect three non-collinear points, or to connect a line and a point, to define a plane. The second axiom stipulates that two planes determine a line, that three non-coaxial planes, or one plane and one line not belonging to the plane, determine a point. The third axiom states that two planes determined by the points A_1, A_2 and A_3, and by the points A_1, A_2, and A_4, define the line A_1A_2. I let the reader interpret the other postulates. Note that the link between projective geometry and incidence relations was not absent in (1897a) (see for instance p. 123: 'the successive application, to any figure, of two reciprocal operations of projection and section, is regarded as producing a figure projectively indistinguishable from the first').

47. With the help of the algebraic version of his postulates, Russell proves that (1899b, p. 407) 'if 1, 2, 3, 4, 5 be five terms such that no two pairs and no two triads have the same product, then the expression 45{23.4(12.345)(13.245)} is constant so long as 145, 245, 345 are constant, however the terms 1, 2, 3, 4, 5 may be changed'; this, explained Russell, amounts to proving Von Staudt's unicity theorem.

48. When the dimension of the embedding space is 3, this case corresponds to the 'projecting' case. For instance, the product of a point and a line (not containing the point) is the plane that contains both.

49. When the dimension of the embedding space is 3, this case corresponds to the 'intersecting' case. For instance, the product of a plane and a line (not in the plane) is the point at which they intersect.

50. I refer the interested reader to Gandon (2004) where I explore in detail the link between Whitehead (1898) and Russell (1899b).

51. Note that Poincaré acknowledged that the '*géométrie de Staudt*' provided an alternative to his own group-theoretical foundation (1898, pp. 28–9). He thought, however, that this path, based on the incidence relations, was very artificial: 'to obtain the notion of length only as a particular case of the anharmonic ratio is an artificial detour one is reluctant to make'. See also pp. 74–5 in this book.

52. Pieri actually translated Von Staudt (1847) into Italian.

53. But see Hallett and Majer (2004) to qualify the standard story.

54. In connecting his considerations of complex numbers and geometry, Russell resumed a trend which went back at least to A. de Morgan (see PoM, pp. 376–7) and which is prominent in Whitehead (1898). For Russell, the point of comparing the complex number fields and geometry is to emphasize that the essence of a space is not the identities of the points but the kind of relations which hold between them. See (PoM, p. 8): 'the mutual relations of points in a Euclidean plane are of the same type as those of the complex numbers; hence plane geometry, considered as a branch of pure mathematics, ought not to decide whether its variables are points or complex numbers or some other sets of entities having the same type of mutual relations'.

55. One finds numerous references to Russell (PoM) in Whitehead's book. What is more, Whitehead (1906b) is in fact the first piece of a two volume work – the first one (1906b) is on projective theory, and maintained the content of PoM chapter 55; the second one (1907) is on descriptive geometry, and extended PoM chapter 56. The two tracts (which were intended to serve as textbooks) maintained the central distinction Russell presented in PoM.

56. On incidence geometries and block designs, see Dembowski (1968) and Buekenhout (1995).

57. In a (v, k, λ)-BIBD, the set of points is finite, which is not necessarily the case in Whitehead's cross-classification. Moreover, in a cross-classification, lines do not

have necessarily the same number of points. Finally, a cross-classification should not be restricted to two-dimensional space.

58. I recount this story in more detail in Gandon (2009a, p. 2831; 2010b).
59. When Pasch's system is endowed with a relevant axiom of continuity, see Note 33 in this chapter.
60. In Gandon (2009b), I argue that this definition of space played a role in Russell's reception of neutral monism. Indeed, his discussion is intimately linked to one about the various kinds of possible classifications. See for instance Russell (1913, p. 15): '"neutral monism" ... is the theory that the things commonly regarded as mental and the things commonly regarded as physical do not differ in respect of any intrinsic property possessed by the one set and not by the other, but differ only in respect of arrangement and context. The theory may be illustrated by comparison with a postal directory, in which the same names come twice over, once in alphabetical and once in geographical order; we may compare the alphabetical order to the mental, and the geographical order to the physical ... Just as every man in the directory has two kinds of neighbours, namely alphabetical neighbours and geographical neighbours, so every object will lie at the intersection of two causal series with different laws, namely the mental series and the physical series.'
61. For more on this approach, see pp. 21–2 and 93.

2   Metrical Geometry

1. On the impact of the emergence of non-Euclidean geometries on mathematics and philosophy, see Torretti (1978) and Voelke (2005).
2. The idea has been continued by Coffa (1981) and Boolos (1994).
3. Musgrave (1977, p. 113): 'I now turn to the logical positivists, and to our historical riddle. The solution to the riddle is this: it was not old-style logicism which the positivists adopted, but rather logicism spiced with varying doses of If-thenism. Mind you, the *rhetoric* of old-style logicism persists, and is used as a stick to beat philosophical opponents.' The riddle, which Musgrave here alludes to and which is introduced at the beginning of his paper, is the following: how can we explain that the logical empiricists made the logicist thesis a cornerstone of their position, while at the same time acknowledging the difficulties which beset the programme?
4. See Coffa (1981), Landini (1998), Grattan-Guinness (2000, p. 303) and Griffin (2002).
5. In chapter 47, Russell examines the case of distance (§§ 396–9) and then the case of angle (§§ 400–4). Here, I leave aside the case of the angle.
6. Klein here presupposes that a distance between points on a real projective line D is a function from $D \times D$ to $\mathbb{R} \cup \{\infty\}$.
7. In the detail, things are more complicated. Thus, I have not here taken into account the determination of the angular distance. For a modernized presentation of Klein's work, see Struve and Struve (2004).
8. Speaking of the Euclidean and hyperbolic cases, Russell says (PoM, pp. 425–6): 'the theory we have hitherto examined, since it used the distinction of real and ideal elements, was descriptive, not projective; we have now to examine the corresponding theory for pure projective geometry. Here there are no ideal elements of the above sort associated with our line.'
9. Let P be the real projective line, and I an involution defined on P. Two points X (x_1, x_2) and Y (y_1, y_2) of P are correlated by the involution I if and only if (*)

$ax_1y_1 + bx_2y_2 + cx_1y_2 + cx_2y_1 = 0$ (the involution I is thus completely characterized by the three real parameters a, b, c, where $ab - c^2$ is not equal to 0). Now, a point (x, y) which is left invariant by I is such that (**) $ax^2 + by^2 + 2cxy = 0$. This last equation has either two real roots or two imaginary conjugate roots (the condition $ab - c^2 \neq 0$ forbids the case of a single root). In both cases, the knowledge of the two roots allows us to determine the coefficient a, b, c occurring on the equations (*) and (**). Note that the (*) expresses that the symmetric bilinear form is equal to 0, while (**) expresses that the associated quadratic form is equal to 0. Thus the passage from the (symmetric) bilinear to the quadratic form encodes the passage from consideration of transformations to consideration of point pairs.

10. Russell (PoM, p. 427): 'involution may be called ideal point-pair: those that correspond to an actual point pair are called *hyperbolic*, the others *elliptic*'.

11. In particular, there is a problem in distinguishing between the two conjugate points. The interested reader can consult Klein (1925, pp. 123–9).

12. Richards (1988, p. 143ff.) shows how important Klein was in the introduction of non-Euclidean geometries in England.

13. At the beginning of his presentation of the projective derivation (PoM, p. 422), Russell refers to a 'fuller discussion than the following' contained 'in my *Foundations of Geometry*, Cambridge, 1897, §§ 30–38'.

14. Klein (1871, p. 72): '[the] three geometries will turn out to be special cases of the general Cayley measure. One obtains the parabolic (ordinary) geometry by letting the fundamental surface for the Cayley measure degenerate to an imaginary conic section. If one takes the fundamental surface to be a proper, but imaginary surface of the second degree, one obtains the elliptic geometry. Finally, the hyperbolic geometry is obtained when one takes the fundamental surface to be a real, but not ruled, surface of the second degree and considers the points inside it'.

15. Klein (1871, pp. 99–100): 'an imaginary point pair, as we have mentioned in passing, may be regarded as a bridge between a real and imaginary conic section, and for this reason parabolic geometry may also be regarded as the transitional case between hyperbolic and elliptic geometry. Suppose, for example, that a hyperbola is given, whose (imaginary) minor axis has a fixed value, while its principal axis gradually shrinks to zero and then becomes imaginary. At the limit zero the two branches of the hyperbola collapse to a doubly-counted line, the minor axis. This line represents the conic section, in so far as it is generated by points. But in so far as it is enveloped by lines, it degenerates to two conjugate imaginary points, which lie at the ends of the constant minor axis on the doubly-counted line.'

16. See also Russell (1897a, pp. 88–9): '[the point I will explain] is of great importance, and underlies, I think, most of the philosophical fallacies of Riemann's school. A judgement of magnitude is always a judgement of comparison, and what is more, the comparison is never concerned with quality, but only with quantity. Quality, in the judgement of magnitude is supposed identical, in the object whose magnitude is stated, and in the unit with which it is compared … It remains to be proved that the comparison, which we can institute between various spaces, is capable of expression in a quantitative form. Rather it would seem that the difference of quality is such as to preclude quantitative comparison between different spaces, and therefore also to preclude all judgements of magnitude about space as a whole. Here an exception might seem to be demanded by non-Euclidean spaces, whose space-constants give a definite magnitude, characterizing space as a magnitude. But this is a mistake. For the space-constant,

in such spaces, is the ultimate unit, the fixed term in all quantitative comparison; it is itself, therefore, destitute of quantity, since there is no independently given magnitude with which to compare it.'

17. One could still object that it is Russell himself, in the 1937 Preface, who spoke of a mutual inconsistency between the geometrical systems as the starting point of his work. It would not be the first time, however, that Russell, coming back to his former writings, misunderstood himself. And one must keep in mind as well that, as I will soon explain, Klein's view is ultimately rejected by Russell. When metrical geometry is viewed as an empirical science, there is a mutual inconsistency between the different geometries, in the sense that they cannot all describe the properties of the physical space. See pp. 66–71 for more on this point.

18. The eight conditions are the following (PoM, p. 408): (1) Every pair of points has one and only one distance; (2) distances are symmetrical relations; (3) on a given straight line through a given point, there are two and only two points at a given distance from the given point; (4) there is no maximum distance; (5) the distance from a point to itself is zero; (6) there is no minimum to the distance between distinct points; (7) the Archimedean axiom; (8) the 'linearity' axiom (according to which a given distance can be divided in n equal parts). Russell was inspired by Klein (1871, § 2), and referred also to Pasch (1882, § 12). Note that these eight conditions do not characterize completely the notion of a distance – they are necessary conditions, which are not however sufficient.

19. According to Russell, it is necessary to take distance as a symmetrical relation for doing plane and spatial geometry (PoM, p. 409): 'as soon as we consider the fact that distances on different [non-parallel] lines may be equal, we see that the difference of sense between AB and BA is not relevant to distance, since there is no such difference between distances on different lines'.

20. Russell did not refer to Pieri (1898), but to Pieri (1900).

21. Peano (1903).

22. Thus, when referring to Pieri (1900), Russell mentions the use of motion (i.e. isometric transformation). It is odd that he did not refer to Poincaré in PoM § 395.

23. On this point, Russell has changed his opinion. In (1901c), he asserted it was not possible to derive descriptive and projective geometry from distance.

24. See PoM (pp. 177–80).

25. The quote mentions descriptive, and not projective, space. As I have explained above, Russell used to consider descriptive space when he dealt with the hyperbolic and Euclidean cases, and projective space when he dealt with the elliptic case. The essential point is however that what works in the descriptive (hyperbolic/Euclidean) case (defining distance as divisibility of a stretch) also works in the projective (elliptic) case. In (PoM, pp. 413–14), he extends what he says here to the projective line.

26. Russell developed a theory of angles as magnitudes of divisibility in PoM §§ 400–3. He did not endorse the view that angle is an empirical concept, though, he did introduce 'some method of connecting the measure of angles with that of distances' (PoM, p. 416). Angle is, thus, for him, a derived quantity.

27. The three further axioms are: the assumption that, on a given line, 'there are two and only two points at a given distance from [a] given point' (PoM, p. 408); a version of the Archimedean axiom; and the assumption that a given stretch can be divided into n equal parts.

28. Russell acknowledged that his theory of magnitude of divisibility was not satisfying in a letter to Couturat dated 13 June 1904: 'today, I believe that I have made a too important use of divisibilities'.

29. Thus, at the time of PoM, Russell shared Poincaré's scepticism about the experiments aimed at determining the curvature of physical spaces by measuring the distances between the Earth and some stellar bodies. He held indeed that the usual optical and mechanical laws used in the measurements presupposed Euclidean geometry (1902, p. 502).

30. Compare with PoM (p. 412): 'the habit of allowing the imagination to dwell upon actual space has made the order of points appear in some way intrinsic or essential, and not merely relative to one of many possible ordering relations. But this point of view is not logical: it arises, in regard to actual space, only from the fact that the generating relations of actual space have a quite peculiar connection with our perceptions.'

31. Poincaré's atomism is reinforced by the idea that each sensation is linked to a particular nerve ending. See for instance the considerations about the 'distribution board' in (1908, pp. 94–6).

32. I thank J. Galaugher for her translation.

33. Note that Russell does not say here that one has an immediate perception of the metric relation itself. He could indeed have entertained the weaker thesis that one has an immediate perception of the instantiated relations of distance, and that, from this perception, one can extract some information on the notion of distance. What is important, however, is that even this weaker thesis contradicts the fundamental tenet of Kant's and Poincaré's theory of perception.

34. See Russell (1904, pp. 439–45). For more on Meinong's, and more generally on the Brentanist's, theory of perception, see Smith (1994).

35. Russell (1904, p. 437): 'a melody of four notes is not a fifth note, and generally a complex is not formed by adding an object to the constituents; nevertheless, something is added. What is added is a relation.'

36. See, for instance, PoM (p. 373): 'the view that no other axioms [than the Euclidean ones] would give results consistent with experience ... could only be tested by a greater mathematical ability than falls to the lot of most philosophers. Accordingly the test was wanting until Lobatchewsky and Bolyai developed their non-Euclidean system. It was then proved, with all the cogency of mathematical demonstration, that premises other than Euclid's could give results empirically indistinguishable, within the limits of observation, from those of the orthodox system.'

37. See also Poincaré (1913, p. 30): 'our body is our first measuring instrument; as with other bodies, it consists of several solid parts, which can move some in relation to others, and certain sensations inform us of the relative displacements of these parts, so that, as in the case of artificial instruments, we know if our body has moved like an invariable solid body'.

38. To show how surprising this choice is, let me recall Russell's famous passage (1914b, p. 112) about Occam's razor: 'the above extrusion of permanent things affords an example of the maxim which inspires all scientific philosophizing, namely "Occam's razor": Entities are not to be multiplied without necessity. In other words, in dealing with any subject-matter, find out what entities are undeniably involved, and state everything in terms of these entities. Very often the resulting statement is more complicated and difficult than one which, like common sense and most philosophy, assumes hypothetical entities whose existence there is no good reason to believe in.' Russell's projective theory of metric is more 'economic' than the divisibility approach, since the latter introduces a new indefinable. Russell, against Occam's razor, opts nevertheless for the

former. I am not saying that Russell contradicts himself – Occam's razor is a principle which does not yet play in 1903 the role it will play later on. But still: the fact that Russell took a decision which goes against Occam's razor is worth noting.

39. Russell emphasized the existence of these ad hoc constraints in his presentation (see PoM, pp. 423–6).

40. For instance, in the 'general' case (that is, the non-Euclidean case), the distance is the product of the logarithm of the cross-ratio by a constant c. The value of this parameter is arbitrarily fixed, and its determination aimed at nothing more than ensuring that the distances considered had real values (Klein 1871, §§ 3, 5); for instance (pp. 81, 82): '[in the hyperbolic case], if we make the requirement that the distance between two neighbouring points on the line be real, we must take the constant c multiplying the logarithm to be real ... [In the elliptic case,] we must ... give c a pure imaginary value ... in order to have a real distance between real points.'

41. (1871, pp. 81–2, 96–9).

42. Notably the constant c was required to grow to infinity (1871, § 6, pp. 83–5).

43. Before PoM, as for instance in Russell (1899e), he sharply distinguished philosophical from mathematical definitions. The way of making the distinction varied according to the texts, but, roughly said, a complex of concepts is a mathematical definition of B if B is uniquely denoted by the complex, while a philosophical definition of B is intended to display the meaning of B. For instance, to characterize 3 as the first odd prime number is a mathematical definition of 3, not a philosophical analysis of the notion. In PoM, Russell downplays this distinction, arguing that in his book only mathematical definitions are important (see for instance p. 112). I do not think however that this change puts an end to the idea that some definitions are more fundamental than some others. On the contrary, we have seen in Chapter 1 and this chapter that Russell went to enormous trouble to select among many formally correct definitions the one which captures the meaning of the concept. In a sense, Klein's projective characterization is a mathematical definition of distance, while Russell's account in terms of divisibility aims at being a philosophical definition of the concept. The problem which is in front of us is then to explain the nature of the difference between the two cases.

44. Torretti (1978, p. 53) tells us that the term had been used by Gauss, when he explained to Bessel that he did not publish anything about his research on non-Euclidean geometry for fear of 'the uproar of Beotians'.

45. Let me quote the letter J. Dieudonné wrote to the French historian of mathematics P. Dugac on 12 May 1984 (Dugac 2003, pp. 221–2): 'the controversy between Poincaré and Russell is very enlightening; it reveals quite obviously how completely *invalid* the reasonings of the alleged "mathematician" Russell are about everything connected to mathematics; he could have been wholly self-taught on the subject, since what he says shows he apparently did not know anything about the works on the foundation of geometry, from Cayley to Pasch, Klein, and the Italian school ... I think that Poincaré was too kind to find time to discuss this verbiage, and to explain the Erlangen Programme again and again ...; this was, however, a waste of effort, since Russell did not understand anything about it ... The moral is: philosophers should know something about mathematics before claiming to speak about it!'

46. See for instance Russell's letter to Jourdain of 1 June 1907 (Grattan-Guinness 1977, p. 105): 'consideration of the paradox of the liar and its analogs has led

me to be chary of treating propositions as entities. I therefore no longer regard as valid the proof of $\aleph_0$ entities. ... I now think that existence-theorems beyond the finite require a definite assumption that the number of entities is not finite.'

3   Geometry, Logicism and 'If-Thenism'

1. Let T be the conjunction of the axioms of a given axiomatized theory, and p one of its theorems. From the deduction theorem, it follows that p derives from T (T $\vdash p$) if and only if T$\Rightarrow p$ is a logical truth ($\vdash$ T $\Rightarrow p$). If-thenism claims that to consider a given theory as a part of logic, it suffices to demonstrate that, T being the axioms of the said theory, the translation of T $\Rightarrow p$ in the logical language is a logical truth.

2. In particular, note that, according to Russell, the axiomatic characterization, for being regarded as a definition of a concept, should not necessarily be categorical. Thus, the relation of order is defined by a list of properties, and there is obviously no isomorphic relation between the various ordinal structures.

3. Russell makes this clear in chapter 49 (PoM, p. 436): 'not only are the actual terms composing a space irrelevant, and only their relations important, but even the relations do not require individual determination, but only specification as members of certain logical classes of relations. These logical classes are the elements used in geometrical definitions, and these are definable in terms of the small collection of indefinables out of which the logical calculus (including that of relations) is built up.'

4. One finds exactly the same criticism in Frege. See for instance his (1903a, pp. 36-7): 'how do things stand with Mr. Hilbert's definition? Apparently every single point is an object. From this it follows that the concept of a point (*is a point*) is of the first level, and consequently that all of its characteristics must be of the first level. If we now go through Hilbert's axioms, considering them as parts of the definition of a point, we find the characteristics stated in them are not of the first level. That is, they are not properties an object must have in order to be a point. Rather they are of the second level. Therefore, if any concept is defined by means of them, it can only be a second-level concept'. For more on this, see Demopoulos (1994).

5. For a discussion, see section 3.4 below.

6. Russell is not perfectly clear on this point, but the idea is certainly to introduce homogeneous coordinates and to define a projective space in this way as a completion of an affine map (see sections 1.2 and 1.6 in Chapter 1).

7. Boolos (1994) has argued that the recourse to an axiom of infinity amounts to giving up logicism. For a discussion of this claim, see Linsky (1999, pp. 89-109) and Landini (2006; 2007).

8. This should be qualified, since Russell and Whitehead took the trouble to develop a large part of arithmetic without resorting to the infinity axiom – by using an ascent in the type hierarchy (see the Prefatory Statement at the beginning of the second volume of PM, and chapter 13 of Russell (1919)). But for developing real analysis, the infinity axiom is absolutely required. Thus, owing to this fact, the status of real analysis is at least here aligned with one of the empirical sciences.

9. See for instance Hylton (1990) and Landini (1998).

10. In the case of projective space, the path followed in the proof of the existence-theorem reverses the path followed in the definitional process: in PoM § 474, the existence of projective space is derived from the existence of Euclidean space,

while in PoM VI, Euclidean geometry (as a pure mathematical theory) is seen as a part of the projective theory. But the notion of a progression illustrates the same pattern. A progression is shown to be not empty by referring to the finite cardinal numbers. One should nevertheless distinguish the concept of a progression from ℕ.

11. See PoM (p. 114): 'thus, so far as the definition by abstraction can show, any set of entities [*a*] to each of which some class has a certain many–one relation [*S*], and to one and only one of which any given class have this relation, and which are such that all classes similar to a given class have this relation to one and the same entity of the set, appear as the set of numbers, and any set is the number of some class'.

12. More precisely, Russell (1901a, p. 321) defined the class *S* of all the many–one relations *S* such that:

$$\forall x \forall y (x \text{ is similar to } y \Leftrightarrow \exists w\, (xSw \,\&\, ySw))$$

To each domain of the relations belonging to this class *S* corresponds a possible set of cardinal numbers – that is, each set of 'properties' is such that if any two classes *x* and *y* have a relation to one of them, then they are similar. See Landini (1998, pp. 24–30; 2006, pp. 228–31).

13. Russell (1901a) states that the principle of abstraction is proved by saying that the class of the classes similar to *u* exists (this would be ruled out in ramified type theory), and thus that there is a relation *S* (of set membership) and an entity *a* (the class of classes similar to *u*) such that all the classes similar to *u* have *S* to *a*. See proof *6.2 in § 1 (p. 320); and, for a comment, Landini (1998, pp. 26–7).

14. This is not so surprising. That the existence-theorems in arithmetic should be built on the very definition of the basic arithmetical concepts seems to follow from the claim that the existence-theorems in mathematics are all obtained from arithmetic.

15. See also p. 133. Concerning the difficulty about getting rid of the psychological idea of a set arranged in his natural order, see p. 242.

16. Frege (1892) made the same kind of distinction. He sharply differentiated *Sinn* (logical content) and *Vorstellung* (psychological images associated with the logical content).

17. See also PoM (p. 47) where he criticized Bradley's psychologism.

18. The young (he was 25 years old in 1905) Boutroux was the son of the French spiritual philosopher Emile Boutroux and married Alice Poincaré, the sister of Henri Poincaré.

19. 'Let us take as an example an important part of mathematical analysis, the study of the differential equations, that is, the search of functions which satisfy certain classes of differential equations. Does the notion of order occur in this study? When we are inquiring whether the integral function has singular points, where they are located, how it behaves in the neighbourhood of these points, in which domain the function can be represented – are we ultimately inquiring, without us knowing it, about the notion of order?' (1905, p. 626).

20. Boutroux (1905) is very close to the criticism developed by Poincaré (1908, p. 125 ff.). Owing to their relationship, it is likely that Poincaré and Boutroux had plenty of occasions to discuss these topics, and it is thus difficult to know who influenced whom.

21. See also Russell (1905b, pp. 263–4): '[Boutroux] shows that the word function is, grammatically, capable of many meanings; this is a fact to interest the lexicographer. He shows, moreover, that the definition of this or that particular function is sometimes difficult. But this does not obscure the meaning of the word

function, any more than a doubt about the identity of the Iron Mask obscures the meaning of the word *man* ... This particular case illustrates, if I am not mistaken, a confusion which vitiates many of the lines of argument of Mr Boutroux, to wit, the confusion between the act of discovery and the proposition discovered. He seems to believe that, because our knowledge of mathematics changes, mathematics itself changes ... The main point is that what evolves is our *knowledge* of mathematics, and not the body of truths which we gradually discover.'

22. See, for other examples of the same move, Russell (PoM, pp. 114, 133) and (1919, pp. 14–16).

23. The if-thenist reading of Russell I am describing here has been developed in Detlefsen (1993).

24. Russell wanted probably to take advantage of the well-known isomorphism between the Euclidean plane and the complex field. The philosopher generalized in PoM VI (pp. 429–30) what he said here about the notion of Euclidean space (quoted on p. 80 above). He also wrote (PoM, p. 436): 'not only are the actual terms composing a space irrelevant, and only their relations important, but even the relations do not require individual determination, but only specification as members of certain logical classes of relations. These logical classes are the elements used in geometrical definitions, and these are definable in terms of the small collection of indefinables out of which the logical calculus (including that of relations) is built up.'

25. See PoM (p. 472): 'before applying these remarks to motion, we must examine the difficult idea of occupying a place at a time. Here again, we seem to have an irreducible triangular relation. If there is to be motion, we must not analyse the relation into occupation of place and occupation of a time. For a moving particle occupies many places, and the essence of motion lies in the fact that they are occupied at different times'.

26. I refer here to the importance the notion of a path had in the development of graph theory – especially the role played by the issues concerning Eulerian and Hamiltonian paths.

27. See the summary Russell gave in PoM (p. 421): 'the true founder of non-quantitative Geometry is Von Staudt ... But there remained one further step, before projective Geometry could be considered complete, and this step was taken by Pieri ... Thus at last the long process by which projective Geometry has purified itself from every metrical taint is completed.'

28. Sinaceur (1991, p. 229) writes: 'one could think that there is a relation between these three pages from Veblen, and Artin and Schreier (1926)', but she adds immediately: 'the reality is very different, however! B. L. van der Waerden, who took part in the seminar sessions where the theory of the real fields was developed, thinks that neither Artin nor Schreier knew of this short article from Veblen'. And she concludes: 'between Veblen's brainwave and Artin and Schreier (1926), the relation is conceptual, not historical. This makes it even more noteworthy.'

29. For more on this, see Corry (1996) and Sinaceur (1991).

30. The Polish logician Adolf Lindenbaum (1935) tried to find some logical criteria to distinguish between formally equivalent theories. In particular, he attempted to develop a general account of the internal simplicity of a concept. See Lindenbaum (1935, p. 32): 'according to a well-known metaphor from Vailati, the deductive systems are ruled by a democratic government: no term is supposed to rise above the others, as the primitive term from which all the others depend. And this is the case, because the choice of what we recognize as primitive terms is in

some way arbitrary ... I would say however that this standpoint of mechanical democracy is already a bit old-fashioned. Time has come for those primitive terms that are in the best position to seize power. One of the criteria (not the unique one) allowing us to choose among these terms is their internal simplicity, or rather the simplicity of a system of terms, since it is usually several independent terms that are jointly put to work.' Lindenbaum attempted to define a logical criterion capable of gauging the simplicity of a concept and a theory – and he wanted to use this notion to restructure the mathematical material. In this last respect, his programme was different from Russell's logicism (and closer to the structuralist reform). Note that, like Russell, Lindenbaum was greatly influenced by Pieri (on this point, see Marchisotto and Smith 2007, pp. 358–62).

4 Quantity in *The Principles of Mathematics*

1. See Rodriguez-Consuegra (1991) and Levine (1998).
2. See Griffin (1991) and Grattan-Guinness (2000).
3. On this, see Dedekind (1888, p. 38): 'all the more beautiful it appears to me that without any notion of measurable quantities and simply by a finite system of simple thought-steps man can advance to the creation of the pure continuous number-domain; and only by this means in my view is it possible for him to render the notion of continuous space clear and definite'.
4. In Gandon (2009e), I compared Russell's 1903 theory of quantity with Meinong's and Poincaré's.
5. In PoM, Russell resumed Dedekind's cut construction and held that the notion of continuity should be severed from any relation to quantity. As I have just said, PoM III is, in this respect, a 'concession to tradition'. In PM, Russell challenged his earlier view and connected real numbers with quantities.
6. One can mention Grassmann (1861), Stolz (1885), Helmholtz (1887), Bettazzi (1890), Poincaré (1893), Veronese (1891), Weber (1895–96), Burali-Forti (1898), Hilbert (1899), Hölder (1901), Huntington (1902) and Hahn (1907).
7. See Ehrlich (2006).
8. The very first definition of the fifth part of the *Elements* presupposes these two ingredients. Euclid said that 'a magnitude is a part of a magnitude, the less of the greater, when it measures the greater'. In other words, B measures A iff $B < A$ and if $A = B + \cdots + B$.
9. The semi-group $\langle A, + \rangle$ is a structure in which $+$ is an associative binary operation defined on A.
10. A copy of Burali-Forti's article, annotated by Russell, can be found in the Russell Archives (McMaster University). Moreover, Whitehead alludes to Burali-Forti in a letter to Russell dated 28 January 1913 (held in the Russell archives in McMaster University): 'as to the preface [to PM VI on quantity] – the work on "grandeurs" started with a study of Burali-Forti's articles in the *Rivista* and was directed initially to arrive at the same results. Of course his work is really based on Euclid Bk V – whom I ought also to have studied, but did not. Thus our antecedents are Euclid and Burali-Forti.' In PoM (p. 181), Russell also referred to Bettazzi.
11. Burali-Forti wrote his axioms in the following way (1898, p. 148): 1. $a + b = b + a$; 2. $a + (b + c) = a + b + c$; 3. $a + b = b + c .\supset. a = b$; 4'. $_{3}G_{0}\text{-}G$; 4''. $_{3}G$; 5. $a\varepsilon G .\supset. a + b\varepsilon G$; 6. $a = b .\cup. a < b. \cup. a > b$; 7. $a\varepsilon G. \supset. _{3}G\cap\theta a$; 8. $u\varepsilon \text{ClsExisLim}'G_{0} .\supset. _{3}G_{0}\cap x\ni(\theta x = \theta u)$. G is here the set of the magnitudes not equal to 0; G_{0} is the union of G and $\{0\}$; θu, where u is a bounded set, is the set of magnitudes which are less than a member of u.

12. The uniqueness of the 0 is a consequence of the first axiom.
13. Axiom 7 says that the ordered set is dense. Axiom 8 says that the ordered set is Dedekind's complete. Burali-Forti does not introduce the Archimedean axiom, but he claims that the Archimedean condition follows from his axioms.
14. The idea that extensive magnitudes are measurable can be logically grounded on the fact that there is a unique increasing homomorphism (modulo the choice of a unity) between magnitude (as defined, for instance, in Burali-Forti) and the real numbers. For more on this, see Hölder (1901) and Krantz et al. (1971). In the nineteenth century, the idea that extensive magnitudes were measurable was considered as too obvious to require a justification.
15. Temperature is not an extensive magnitude. For a history of the measurement of temperature and an overview of the debates it generated, see Chang (2004).
16. Many books have been written on the discussion surrounding the emergence of measurement in psychophysics (but see Heidelberger, 2004; and Stevens, 1986).
17. This contrasts a lot with the situation one finds today in measurement theory. In Suppes's approach, measurement is viewed as a numerical representation of an empirical relational structure. The existence of a measurement is thus linked with the possibility of proving a representation theorem. In this perspective, extensive magnitudes are just one among many various measurement systems. Suppes and Zinnes (1968) is a good introduction to measurement theory; for more, see Krantz et al. (1971). For a history of the movement, see Diez (1997).
18. See, for instance, PoM (pp. 160–1): 'some quantities are indivisible. For it is generally admitted that some psychical existents, such as pleasure and pain, are quantitative ... There is ... no reason to regard pleasures as consisting of definite sums of indivisible units.'
19. At the end of chapter 19, Russell characterizes the concept of a magnitude by these following conditions (pp. 168–9): '(1) Every magnitude has to some term the relation which makes it of a certain kind. (2) Any two magnitudes of the same kind are one greater and the other less. (3) Two magnitudes of the same kind, if capable of occupying space or time, cannot both have the same spatio-temporal position; if relations, can never be both relations between the same pairs of terms. (4) No magnitude is greater than itself. (5) If *A* is greater than *B*, *B* is less than *A*, and vice versa. (6) If *A* is greater than *B* and *B* is greater than *C*, then *A* is greater than *C*.'
20. Note how various these kinds of magnitude are: Russell mentioned some mathematical kinds (length, area, volume, angle, etc.), but he also spoke of mass, luminosity, price, pleasure, suffering, resemblance to a sample, etc.
21. See Russell's discussion (PoM, p. 174) of Bentham's slogan according to which 'quantity of pleasure being equal, pushpin is as good as poetry'.
22. See PoM (p. 171): 'the difference or resemblance of two colours is a relation, and is a magnitude; for it is greater or less than other differences or resemblances'. See also pp. 161–2.
23. Russell claimed that relational quantities cannot be divided. This is not surprising. One can understand that a spatio-temporal whole is divisible, but how can one divide a relation or an instantiated relation? To see how a relational magnitude can be measured, see pp. 126–32.
24. On the relation of arithmetical addition with set-theoretical union, see PoM Chapter 12 (especially p. 118): 'if *k* be a class of classes no two of which have any common terms (called for short an exclusive class of classes), then the arithmetical sum of the numbers of the various class *k* is the number of terms in the logical sum of *k*'.

25. See PoM (p. 151): 'there is no doubt that the notion of half a league, or half a day, is a legitimate notion. It is therefore necessary to find some sense for fractions in which they do not essentially depend upon number. For, if a given period of twenty-four hours is to be divided into two continuous portions, each of which is to be half of the whole period, there is only one way of doing this: but Cantor has shown that every possible way of dividing the period into two continuous portions divides it into two portions having the same *number* of terms'. See also p. 161.

26. The idea is roughly to view the quantitative wholes as ordered sets, and to define the concatenation operation as an addition of two series which have no term in common (see PoM, p. 321): 'if the fields of P and Q [P and Q being any relation] have no common terms, P + Q is defined to be P or Q or the relation which holds between any term of P and any term of Q, and between no other terms'. This does not work either because, as Russell himself remarks (PoM, p. 161): 'the cardinal number of parts in any two continuous portions of space is the same, as we know from Cantor; even the ordinal number or type is the same for any two lengths whatever'.

27. In particular, Jordan (1882) and Peano (1887). On the history of the theory of measure, see Hawkins (1970) and Michel (1992).

28. Independently of the issue concerning magnitude of divisibility, this absence of any reference to the theory of measure seems to me very puzzling. The only way I can explain this fact is that, at the end of the nineteenth century, mathematicians encountered some difficulty in differentiating the cardinality and the measure of a set. The construction of a nowhere dense set with a positive measure (the Cantor set) helped to clear the situation (for more on this, see Ferreirós (1999, pp. 157–69) and Bressoud (2008, pp. 1–119)). Russell might have believed that measure theory could be reduced to the theory of cardinals or of ordinals. If so, he was mistaken.

29. See also PoM IV, chapter 31, p. 253.

30. On Meinong (1896), see Potrc and Vospernik (1996) and Guigon (2006).

31. Russell devotes the 'Note to Chapter XIX' (PoM, pp. 168–9) to Meinong's theory. Let me quote the beginning: 'the work of Herr Meinong on Weber's Law, already alluded to, is one from which I have learnt so much, and which I so largely agree, that it seems desirable to justify myself on the points in which I depart from it'.

32. In Gandon (2006a), I hold that the insight according to which certain magnitudes are relational comes from Russell's reception of Hegel's dialectic of quantity, which was then very popular among the British Idealists. In Hegel, the notion of a comparison allowed a dynamical synthesis (*überhebung*) between the contradictory concepts of quantity and quality. Thus, Russell (1897c, p. 79) defended the claim that 'a quantity is really as improper an expression for things which can be quantitatively compared, as a "likeness" for a photograph'. This first connection between magnitude and relation preceded the reading of Meinong (1896) and could explain why Russell was so interested in this book.

33. On several occasions, Russell suggested that the case of geometrical distance (the case philosophers usually focused on when they tried to understand the nature of geometrical quantity) is the most liable to cause confusion in thought, because, on the straight line, the distinction between relational distance and a divisible length is more difficult to make than in the case of angles (which are plainly relational) or in the case of areas and volumes (which are plainly divisible). See for instance PoM (p. 182): 'on the straight line ... we have two philosophically

distinct but practically conjoined magnitudes, namely the distance, and the divisibility of the stretch. The former is similar to angles; the latter to areas and volumes ... Owing to the confusion of the two kinds of magnitude connected with the line, either angles, or else areas and volumes, are usually incompatible with the philosophy invented to suit the lines.'

34. Many scholars have stressed this hiatus. See for instance Griffin (1990, pp. 267–9), Rodriguez-Consuegra (1991, pp. 164–5, 189–205) and Levine (1998).

35. See also p. 121: '[6. 2] affirms that all relations which are transitive, symmetrical, and non-null can be analysed as products of a many–one relation and its converse, and the demonstration gives a way in which we are able to do this, without proving that there are not always other ways of doing it. P6. 2 is presupposed in the definitions by abstraction, and it shows that in general these definitions do not give a single individual but a class, since the class of relation S is not in general an element. For each relation S of this class, and for all terms x of R, there is an individual that the definition by abstraction indicates; but the other relations S of that class do not in general give the same individual.'

36. See Byrd (1987).

37. Levine (1998, p. 118): 'the central topics of PoM – the nature of number, magnitude, order, space, time, and matter in motion – constitute Russell's main philosophical interests from R_1 on. In R_1, Russell conceived of these topics as following the dialectical transitions from the more abstract to the more concrete sciences. In R_2, he conceived of them as exhibiting different independent series, all of which exemplify the absolute theory of order. Early in R_3, Russell has no intention of changing his basic outlook of R_2. In his first R_3 draft of PoM, written in November–December 1900, Russell nowhere departs from his commitment to absolute theories of order. However, once he defends the logicist definitions of integers, philosophically as well as mathematically, Russell can no longer maintain that the absolute theory of order is correct in all cases of series. For, as I have discussed above, the logicist definitions of integers amount to a relative theory of integers ... Russell does not, however, make the defence of relative theories of order the central theme of PoM. Perhaps he was not yet willing to abandon fully his R_2 conception of the mathematical sciences ... Perhaps, as these letters suggest, Russell was simply so anxious to complete PoM that he was in no position to undertake the drastic revisions needed to make relative theories of order his overriding theme. However, for whatever reason, Russell does not make either absolute or relative theories of order his central theme in PoM.'

38. Let me quote the famous verse from Verlaine's *Art Poétique*:

> For we desire Nuance yet more –
> Not colour, nothing but Nuance!
> Oh! only nuance brings
> Dream to dream and flute to horn!

39. William Ernest Johnson (1858–1931) was Sidgwick lecturer at Cambridge from 1902 to 1924. He was thus the colleague of Whitehead and Russell. Even if Johnson was a member of the British logic 'old guard', pushed aside by PM, his influence was not empty. He had J. Maynard Keynes as a student, and C. D. Broad and G. F. Stout as opponents in many discussions. For more on Johnson, see Prior (1949) and Poli (2004).

40. See Johnson (1921, pp. 177–8): 'to illustrate more precisely what is meant by "generates"; let us take the determinable "less than 4"; then "less than 4" generates

"3" and "2" and "1" in the sense that the understanding of the meaning of the former carries with it the notion of the latter. Now no substantive class-name generates its members in this way; take, for instance, "the apostles of Jesus", the understanding of this class-name carries with it the notion "men summoned by Jesus to follow him", but it does not generate "Peter and John and James and Matthew etc.", and this fact constitutes one important difference between the relation of sub-determinate to super-determinate adjectives and that of general to singular substantives.'

41. Thus, the analysis of 'the relation of a magnitude to that whose magnitude it is' is presented in chapter 19 as an objection to the absolutist view (see p. 164).

42. This is the meaning of the last clause $\pi = \rho$. The two Greek letters π and ρ designate the domains of (respectively) P and R.

43. In the general case, the relative product can be defined in this way: Let $R \subset X \times Y$ and $S \subset V \times W$, then $RS =_{def} \{(x, z) \in X \times W : \exists y \in Y \cap V, (x, y) \in R \wedge (y, z) \in S\}$. If $Y \cap V = \varnothing$, RS is then the empty relation. For more on the relative product, see PM *34.

44. In fact, to guarantee that one–one relations are 'onto', one needs clause 1 (if P belongs to G, then P^{-1}, being in G, is a one–one relation of domain **D**).

45. $\langle S, + \rangle$ is a Group if and only if (1) $\forall x \forall y \in S, x+y \in S$; (2) there is an i (identity element) in S such that $\forall x \in S, x + i = i + x = x$; (3) $\forall x \exists y\, x + y = y + x = i$.

46. This distinction of three levels is the same as the one expounded at the end of section 4.1. Ground level corresponds to the stage of quantities (the stage where the relational magnitudes are applied to the couples of **D** $\times$ **D**); level 1 is the stage of magnitude; level 2 is the stage of the relations between magnitudes (where order and addition, i.e. relative product, are defined).

47. Russell added a note to 7*1.1 (1900b, p. 609): 'this is a definition of a kind of distance, i.e. of a class of distances which are quantitatively comparable. A kind of distance is a series in which there is a term between any two, and it is also a group. If any two terms belong to the field of this group, there is a relation of the group which holds between them. If Q be the relation in virtue of which the relations of the group form a series, and if R_1, R_2, be relations in the group such that $R_1 Q R_2$, then $R_1 R_2 = R_2 R_1$, and the relation still holds when both sides are multiplied by any other relation of the group.'

48. In fact, Russell needs more – the action must be transitive and 'faithful': the relation R holding between any couple of **D** must exist and be unique.

49. For a symbolic version, see Russell (1900b, p. 610): '$R, R' \in L \supset_{R, R'} \exists n \cap \rho \ni (R^\rho Q R')$'. In this formula, n is a whole number.

50. For a symbolic version, see (1900b, p. 611): '$n \in N \supset_{x,y} \exists \lambda \cap z \ni \{n.(xz) = (xy)\}$'. That is, it is possible to divide any segment (xy) into a segment (xz) such that $n.(xz) = (xy)$.

51. The two systems differ also in two less important respects: Burali-Forti's magnitude is the positive cone of an ordered group (while Russell's distance corresponds to the whole group structure), and the order relation of a Russellian distance does not necessarily satisfy the Dedekindian postulate (axiom 8).

52. Thus, in PoM chapter 21, Russell seems to follow more closely Burali-Forti's axiomatic than in Russell (1900b).

53. In PoM chapter 31 (see p. 253), Russell defines distance as a densely ordered transformation group, in a way which satisfies the conditions set out in Russell (1900b).

54. The only exception I know of is Frege's (1903b) construction of quantitative domain. For more on Frege's conception, see Dummett (1995) and pp. 138–41 in this book.

55. Using the terminology of measurement theory, this would amount to proving a representation theorem.
56. One finds the same degree of precision in Burali-Forti (1898) for instance.
57. In the terminology of measurement theory, Russell shows how one can induce from the extensive scale defined on Δ an interval scale defined on **D**.
58. See footnote 48 p. 128. Strangely enough, that the action of Δ on **D** should be faithful is required in the passage quoted on pp. 117–18 (PoM, p. 180): 'I shall mean by a kind of distance a set of quantitative asymmetrical relations of which *one and only one* holds between any pair of terms of a given class' (my italics).
59. Russell (in 1900b as in 1903) supplements the theory of distance by another doctrine, which could be applied to measurable 'cyclic' magnitudes, where 'cyclic' magnitudes designate some magnitudes which are arranged in a closed series, like angles in Euclidean geometry. A sketch of Russell's theory is presented in § 8 (1900b). Here is the introductory note (pp. 610–11): 'the following is a general theory of which the angles in a plane about a point afford an illustration. We have seen that, in series having a term between any two, it is in general necessary to regard the generating relation as transitive ... But where the generating relation is transitive, it follows, since it is an aliorelative, that the series cannot be closed. Thus none of our previous theories will apply to angles. By considering series in which (as in the case of angles) the terms of the series are themselves asymmetrical relations, this want can be supplied.'
60. I refer the interested reader to Gandon (2009e, pp. 115–17). Note that Russell's reformulation of Von Staudt's construction is certainly influenced by Klein (1871), which defined a 'skala' in terms of projective transformations.
61. The idea of connecting the two questions is not peculiar to Russell, however. It is widely shared by nearly all the mathematicians interested in the topic. The second part of Hölder (1901) is devoted to the introduction of coordinates on the Euclidean line, and even Hilbert's (1899, ch. 3) calculus of segments illustrates the connection between the abstract characterization of a quantitative structure (Hilbert called it a 'number system') and coordination. Once again, what is original in Russell is the idea that distances are relations (transformations).
62. Note that Johnson (1922) devoted a whole chapter of the second volume of his *Logic* to the PoM theory of quantity. A direct influence is thus not excluded.
63. Bigelow (1988) is an attempt to revive this relational theory of magnitude.
64. See PoM chapters 35 and 36. For instance (p. 296): 'the definition of continuity which we examined in the preceding chapter was, as we saw, not purely ordinal; it demanded, in at least two points, some reference to either numbers, or numerically measurable magnitudes. Nevertheless continuity seems like a purely ordinal notion; and this has led Cantor to construct a definition which is free from all elements extraneous to order.'

5   Quantity in *Principia Mathematica*

1. One finds some very brief remarks in Quine (1941; 1963) and Bigelow (1988).
2. I will here leave aside Part D, in which Russell and Whitehead discussed measurement in cyclic magnitude (cyclic vector-family). For more on this, see Gandon (2009e).
3. I will not mention Cantor's and Weierstrass's constructions, because they do not play the role Dedekind's definition played in Russell's thought.
4. See PoM (pp. 274–5).

5. See PoM (p. 286): '[my theory] requires no new axiom, for if there are rationals, there must be segments of rationals; and it removes what seems, mathematically, a wholly unnecessary complication, since, if segments will do all that is required of irrationals, it seems superfluous to introduce a new parallel series with precisely the same mathematical properties.'

6. Burali-Forti was very careful not to introduce numbers in the formulation of his axiomatic system. He explained that he avoided the Archimedean axiom precisely because its formulation would have compelled him to use natural numbers.

7. In the *Elements* V, 5, Euclid claimed that two couples of quantities (q_1, q_2) and (q_3, q_4) are in proportion if and only if $\forall m,n \in \mathbb{N}$ $(mq_1 < nq_2$ & $mq_3 < nq_4)$ or $(mq_1 = nq_2$ & $mq_3 = nq_4)$ or $(mq_1 > nq_2$ & $mq_3 > nq_4)$. The case '$mq_1 = nq_2$ & $mq_3 = nq_4$' corresponds to the case of rational numbers; the two others correspond to the two conditions set out by Burali-Forti in terms of segments of rationals. Note also that this return to the 'old-fashioned algebra of quantities' is part of a larger programme, based on the idea that the central place we usually give to numbers prevents us from understanding the generality of algebra – on this 'Grassmannian' context, see Burali-Forti (1897), and for more on this background, see Gandon (2010a).

8. This is perhaps to block this obvious objection that, after having defined the reals, Burali-Forti showed that the set of positive real numbers endowed with the usual addition satisfies the axioms of a homogeneous magnitude. But if so, then the cure is worse than the disease: as the properties of the real numbers depend on the existence of the structure 'homogeneous magnitude', it is a mistake to prove that the theory has a model by referring to $\langle \mathbb{R}, + \rangle$.

9. More recently, Bigelow (1988) is a development of such a physicalist approach.

10. Hale's goal is to extend the neo-Fregean programme to real analysis. This attempt has been discussed by Shapiro (2000), Wright (2000) and Batitsky (2002). For more on this debate, see Chapter 6.

11. Hale's approach deviates from Frege's in two respects. First, Frege speaks about groups, where Hale's complete q-domain corresponds to the positive cone of a group. Second, Hale defines quantities as an abstract structure, while Frege regards quantities as injective mappings of a set onto itself. This second difference is more important than the first one. For more on this, see footnote 43 below.

12. Frege resorted to the theory of series to prove the existence of complete q-domains (see Simons 1987 and Dummett 1991).

13. Here, the type constraints are left implicit. But of course, in the following definition, if x and y are object variables, R and S must be relations which take objects as arguments.

14. The stress put on the irreducible ratio comes from the wish to avoid any unnecessary ascent in the type hierarchy; see footnote 17 below for more on this topic.

15. Note that, as no hypothesis on R is made, the relational product can no longer be defined as a composition of applications. Note also that, in PM (1913), relations are considered to be incomplete symbols, which adds to the complication of the concept of relational product; see Landini (2007, pp. 178–82).

16. That the order is dense is proved in *304.3. That the product is commutative and associative is proved in *305.11 and *305.41. That the addition is commutative and associative is shown in *306.11 and *306.31.

17. In fact, the situation is even worse than that: the relational feature of *303.01, far from being neutral with respect to the derivation of the arithmetical properties, represented an obstacle that Russell and Whitehead have to overcome. Indeed, let us suppose that m, n, p, q are very big positive integers, that both m, n and

p, q are relative primes, and, furthermore, that $m/n \neq p/q$ (that is, $mq \neq np$). Let us assume also that there are only a finite number of individuals in the universe (i.e. that the axiom of infinity is false). According to the relational part of *303.01, to say that the two relations R and S have the ratio m/n is to say that there are two individuals x and y such that $xR^n y$ and $xS^m y$; in the same way, to say that two relations $T(p/q)U$ is to say that there are two individuals x and y such that $xU^p y$ and $xT^q y$. Now, if the axiom of infinity is not admitted, it might happen that no such couples of relations exist. Recall that, in PM, relations are taken in extension – if only a finite number of individuals exists, then there could not be enough relations to satisfy the conditions $xR^n y$ and $xS^m y$, and $xU^p y$ and $xT^q y$. If this is the case, m/n and p/q should be considered as equal, since they would apply to exactly the same couples of relations – namely: none! In PM, the infinity axiom is thus needed to guarantee that the relational part of *303.01 does not do any harm. It is needed to ensure that one always has enough relations to distinguish between two ratios m/n and p/q as 'arithmetically' distinct. Russell and Whitehead postponed the use of the infinity axiom as long as they could, by relying on the ascent of type strategy (see Prefatory Statement in Volume 2 of PM). However, they needed it to prove the density of the set of ratios.

18. In (1913, pp. 261–2), the following illustration is given. Let's suppose that two sequences of entities P and Q are such that their first term P_1 and Q_1 are the same ($P_1 = Q_1$), that the fifth term of P is the same as the sixth term of Q ($P_5 = Q_6$), that $P_{11} = Q_9$ and that $P_{13} = Q_{25}$, and that P and Q do not contain any other terms in common. Let's call R the relation that any element of P has with its immediate successor, and S the same relation with respect to Q. Then, by construction, $P_1 R^4 P_5$ & $P_1 S^5 P_5$, then $R(5/4)S$. But we also have $P_1 R^{10} P_{11}$ & $P_1 S^8 P_9$, hence $R(4/5)S$; and $P_1 R^{12} P_{13}$ & $P_1 S^{24} P_{25}$, thus $R(2/1)S$. The two relations R, S then have different ratios. Moreover, even if $R(5/4)S$, it is just not true that $R^5 = S^4$.

19. As Quine aptly summarized it (1963, p. 129): 'Whitehead and Russell (*310, *314) systematically developed two versions of the real numbers. In one version … each real number x became, intuitively speaking, the class z of all ratios $< x$. In the other version x became rather Uz; thus intuitively speaking, the union of all ratios $<x$.'

20. $U(X)S$ means (intuitively speaking) that the couple (U, S) has a ratio $<X$. $U(Y)S$ means that (U, S) has a ratio $<Y$; now, from $X \leqslant Y$, one can deduce that $U(X)S$ implies $U(Y)S$.

21. A representation theorem about semigroups (a kind of extension of Cayley's theorem) says that every abstract semigroup can be represented as a semigroup of injective mappings; see Howie (1995).

22. See Howie (1995) for an account of the various representation theorems available in the theory of the semigroup.

23. The point a in α is connected in κ iff $\forall x \in \alpha$, $\exists R \in \kappa$, $R'x = a$.

24. This result plays a crucial role with respect to the distinction between quantity and magnitude. Recall that in PoM, a relational quantity is the instantiation of a relational magnitude. In PM, Russell and Whitehead do not use the old distinction at all – and for good reasons: in the general case, it is not true that vectors partition the Cartesian product $\alpha \times \alpha$. Now, according to *331.43, when κ is connected, then the vectors of κ_l could be regarded as magnitudes, and their instantiations could be viewed as quantities: two elements of α are related by one and only one vector – in other words, there is a many–one relation between $\alpha \times \alpha$ and κ_l. For more on the distinction between magnitude and quantity in the context of PM VI, see Gandon (2009e).

25. An open (connected) family can be illustrated as the set of right-oriented translations of the Euclidean line. Russell (1900b and 1903) distinguished this case from the case of the circle endowed with the right-handed rotations. This opposition gave rise to the distinction between distances and angles, and the contrast is retained in PM: open families are opposed to cyclic families, whose theory is expounded in section D. On the PM doctrine of cyclic family, see Gandon (2009e).

26. A connected family should not be confused with a connected relation. A relation R is connected when any two elements of its field are linked by R or by R^{-1} (1912, p. 516).

27. An α-point a is transitive if and only if $\forall R, S \in \kappa, \exists T \in \kappa, R'S'a = T'a$.

28. A Dedekindian series is one such that each subset of its field has either a maximum or a lower upper bound. $\langle \mathbb{R}, < \rangle$ is then not Dedekindian, since there are subsets of $\mathbb{R}$ which admit neither a maximum nor a lower upper bound (all the not-bounded-above subsets of $\mathbb{R}$). To get back to the usual case, Russell and Whitehead introduce the concept of semi-Dedekindian. A semi-Dedekindian series is one which 'becomes Dedekindian by the addition of one term at the end' (1912, p. 685). Thus, $\langle \mathbb{R}, < \rangle$ is semi-Dedekindian.

29. In particular, Russell and Whitehead's refined use of the theory of group and semigroup action deserves more attention. For more on the history of the theory of the semigroup, see Hollings (2009).

30. In order for κ to be the positive cone of a distance, a continuity assumption (the series should be semi-Dedekindian) and a closure property (κ must contain all the powers of their members) should be added.

31. It is important to note that neither condition (1) nor condition (2) implies that the mapping is 'onto'. That is, it could be the case that, in a given family κ which fulfils the two conditions, there are couples vectors which are not correlated to any positive rational number. This remark is of course important with respect to the issue concerning measurement by real number.

32. Using Suppes's terminology (see Krantz et al. 1971), one could say that proving conditions (1), (2) and (4) amounts to proving a representation theorem for a (rational) extensive scale, and that proving (3) amounts to proving a uniqueness theorem (to 'change the unit' one must multiply all the measures by a fixed rational number, i.e. by the measure of the old unit in the new system).

33. Recall that Russell here showed that to each rational exponent of the unit corresponds a unique distance.

34. Do not forget that PM VI was supposed to be followed by a book entirely devoted to geometry. The references to this geometrical setting are frequent in PM VI; see for example (1913, p. 436): 'the subject of "rational nets" ... is of importance for the introduction of coordinates in geometry'. Recall also that in the letter he wrote to Russell on 14 September 1909, Whitehead connected the theory of quantity to the introduction of metric, since he said (just after having held that 'the old fashioned algebras which talked of "quantities" were right, if they had only known what "quantities" were') that 'the connection between analysis of metrical geometry is immediate – is in fact the same thing'.

35. To illustrate this development, let me take the example of the Euclidean line endowed with the usual coordinates. Let's assume that S is the right translation defined on the Euclidean line by the unit vector (which sends the point origin O to the point of coordinate 1). In this context, the set α would be the Euclidean line, and the class of translations would correspond to the family κ. Now, the (positive) rational multiples of the translation S would constitute a rational family κ'. It is

clear that certain points cannot be reached from the point O through the rational multiples of S – for instance, the point of coordinate $\sqrt{2}$ cannot. The (κ, S, O)-rational net is the restriction of κ' to the points accessible from O (to the points which have rational coordinates).

36. Veblen and Young (1910) introduced the concept of a net of rationality (see pp. 141–68) in a very general projective setting – without specifying whether the underlying projective space is real, complex or finite. In the first chapter of volume 2 (Velden and Young 1914), the different cases are separated by setting some constraints on the relations between the projective line and its nets: (p. 9): 'throughout Vol I we left the character of [the] net indeterminate. It might contain only a finite number of points or it might contain an infinite number. We propose now to introduce a new assumption which will fix definitely the structure of a net of rationality.' It would be worthwhile developing in more detail the comparison between PM VI and Veblen and Young (1910; 1914). The non-publication of the fourth volume of PM makes the development of such a parallel difficult.

37. They recall at this level Klein's criticism: without a continuity assumption (the semi-Dedekindian condition), it is not possible to introduce real coordinates.

38. PM VI was supposed to be a transitional part which should have led us to geometry. Since the last volume of PM was never published, one can only conjecture about the content of the PM account of geometrical space. This loss has also some consequences concerning the interpretation of the published part of PM. For instance, one does not know how Russell and Whitehead planned to use measurements in the following parts – and this makes the interpretation of PM VI very difficult.

39. For Hale, a complete q-domain should be viewed as a paradigm of all scales of measurement whatever. See (2002, pp. 312, 316): 'in essence, I would argue that for the purposes of an explanation of measurement applications of the real numbers, simple extensive quantities such as mass, length, etc., are the fundamental case'.

40. One finds such a generalization of PM's vector-family in Wiener (1921). For a comment on Wiener, see Gandon (2009e).

41. See the conclusion in Shapiro (2000).

42. See Dummett (1995, p. 390): 'Frege is so anxious to press on to his definition of real numbers that he ignores all quantitative domains save those that have the structure of the real line; as a result he offers a highly defective analysis of the concept on which he fastens so much attention'.

43. As I have said, one finds in the *Grundgesetze* the idea that quantities are relations. More exactly, Frege defined quantities as permutations on a domain, and quantitative addition as a composition between permutations. But Frege, following Gauss, only adopted this point of view to explain the nature of negative quantity. His account is then not as rich as the one developed in PM. For more on this, see Dummett (1995, pp. 387–90).

44. For more on structuralism and on objects as places in the structure, see Shapiro (1997, pp. 77–84).

45. Using the terminology of the theory of measurement, one could say that Russell and Whitehead construe extensive scale as a difference scale whose origin is fixed.

6 Application Constraint in *Principia Mathematica*

1. See Desmet (2010, ch. 3) for more on this and Whitehead's conception of physical science.

2. See as well the beginning of *Process and Reality* (1928, p. 6): 'it is a remarkable characteristic of the history of thought that branches of mathematics, developed under the pure imaginative impulse ... finally receive their important application. Time may be wanted. Conic sections had to wait for eighteen hundred years. In more recent years, the theory of probability, the theory of tensors, the theory of matrices are cases in point.'

3. Desmet (2010, pp. 164–6) holds that Whitehead's position is different from Wigner's. His idea is that Whitehead's theory of abstraction provides us with the means to explain the relation between the mathematical concepts and the real world. But it is one thing to explain how abstract concept in general can be applied and another to account for the fact that the particular abstract concepts elaborated by the pure mathematicians are precisely those that are needed by the physicists. This is the second issue I emphasize here. And I do not see any way in which Whitehead's theory of abstraction could solve it.

4. Note that the examples given by Wigner resemble the ones brought forward by Whitehead – complex numbers, here and there, seemed to play a decisive role. Thus Wigner (1960, p. 5): 'let us not forget that the Hilbert space of quantum mechanics is the complex Hilbert space, with a Hermitean scalar product. Surely to the unpreoccupied mind, complex numbers are far from natural or simple and they cannot be suggested by physical observations. Furthermore, the use of complex numbers is in this case not a calculational trick of applied mathematics but comes close to being a necessity in the formulation of the laws of quantum mechanics. Finally, it now begins to appear that not only complex numbers but so-called analytic functions are destined to play a decisive role in the formulation of quantum theory. I am referring to the rapidly developing theory of dispersion relations.' On Wigner's account, see also Steiner (1998).

5. For more on this topic, see Gandon (2005a).

6. For an accessible presentation of Grassmann's algebra, see Grassmann (1862), Peano (1888) and Crowe (1967).

7. Grassmann (1862, p. 126): '[Theorem] 222. The sum $\alpha A + \beta B + ...$, in which $A, B,...$ are points, $\alpha, \beta...$ numbers, is a displacement or a ... point according as $\alpha + \beta ...$ is zero or not.'

8. Ibid. (p. 132): 'in order to be able to handle all the spatial magnitudes of first order, that is the ... points and displacements (of given length and direction), in the same way, I will say that the *position of a displacement* is the infinitely distant point that the lines parallel to this displacement have in common, or alternatively that that displacement is a magnitude of first order that lies on these lines at an infinite distance.'

9. For more on this move, see (1898, pp. 162–5).

10. See especially (1898, pp. 364–7).

11. Let me quote the historical remark which concludes Chapter 1 of book VI (1898, p. 370): 'the idea of starting a "pure" Metrical Geometry with a series of definitions referring to a Positional Manifold is obscurely present in Cayley ... it is explicitly worked out by Homersham Cox ... and Sir R. S. Ball .... Sir R. S. Ball confines himself to three dimensions, and uses Grassmann's idea of the addition of points, but uses none of Grassmann's formulae for multiplication. But the general idea of a pure science of extension, founded upon conventional definitions, which shall include as a special case the geometry of ordinary experience, is clearly stated in Grassmann's *Ausdehnungslehre von 1844.*'

12. As is well known, Russell insisted on the separation between pure and applied mathematics. See for instance (1907a, p. 86): 'mathematics, rightly viewed, possesses not

only truth, but supreme beauty – a beauty cold and austere, like that of sculpture, without appeal to any part of our weaker nature, without the gorgeous trappings of painting or music, yet sublimely pure, and capable of a stern perfection such as only the greatest art can show ... Remote from human passions, remote even from the pitiful facts of nature, the generations have gradually created an ordered cosmos, where pure thought can dwell as in its natural home, and where one, at least, of our nobler impulses can escape from the dreary exile of the actual world.'

13. See for instance the presentation of Cantor's set theory one finds in Baire and Schoenflies (1909). In their presentation, Baire and Schoenflies (in fact, Baire should be considered as the only author of the paper, since he completely reworked Schoenflies original paper written in German) brought forward the various applications of Cantor's doctrine to the theory of the function of the real variable. In a letter to Borel (24 March 1905), he however confessed (Baire 1990, p. 83): '[in the article for Molk's *Encyclopedia*], I have not the right to wander from Cantor's standpoint. Willy-nilly, I have to speak about addition and multiplication of ordinal types, etc., stuff which I do not know any application of. I can hardly invent one! Understand it who may.'

14. Of course, this assessment concerns only the organization of the mathematical material and leaves aside all the logical aspects (the ramified type theory, the theory of incomplete symbols). Note that the need to deal with the contradictions greatly complicates the development of arithmetic, since Russell and Whitehead attempt to develop a large part of arithmetic without resorting to the infinity axiom, by using an ascent in the type hierarchy strategy (see the Prefatory Statement at the beginning of the second volume of PM, and Chapter 13 of Russell (1919)). In saying that there is a correspondence between the organization of PoM and PM, I am just focusing on the shaping of the mathematical material, not on the detailed logical treatment of the various mathematical branches.

15. A series is said to be considered 'rational' if and only if it 'is compact, has no beginning or end, and has $\aleph_0$ terms in its field' (1913, p. 199).

16. A series P is 'continuous' in the sense of Cantor if and only if each subset of its field has a lowest upper bound, and if it contains a denumerable subseries dense in P. A subset which is dense in the field of a series is called by Russell and Whitehead a median class – see *271 (1913, pp. 186–7). And Russell and Whitehead call Dedekindian a series set which is such that each of its subsets has a lowest upper bound. In PM terminology, a continuous series is thus 'a series which is Dedekindian and contains [a denumerable class] as a median class' (1912, p. 577).

17. PoM (p. 298): 'given [a series S of type η], construct all the segments defined by fundamental series in S. These form a perfect series, and between any two terms of the series of segments there is a segment whose upper (or lower) limit is a term of S. Segments of this kind, which may be called rational segments, are a series of the same type as S, and are contained in the whole series of segments in the required manner.' Russell closely follows Cantor. In particular, the notions of fundamental series and of perfect series directly come from Cantor (1895–97). The ordinal definition of a fundamental series is given in PoM (pp. 283–4) – and the definition of a perfect series is expounded in PoM (p. 297). In PM, the notion of a perfect series is replaced by the notion of a Dedekindian series.

18. See Russell (1919, p. 9): 'that "0" and "number" and "successor" cannot be defined by means of Peano's five axioms ... is important. We want our numbers not merely to verify mathematical formulae, but to apply in the right way to common objects. We want to have ten fingers and one nose.'

19. The plan of PM V is as follows. Section A is devoted to the definitions of the basic elements of the theory of series (transitive relation *201, connected relations *202, etc.) and to the construction of the concept of limit (*207) and lower upper bound (called sequent, *206). In section B, Russell and Whitehead review the fundamentals of Dedekind's and Cantor's theory: the concepts of cut (segment, *211, *212), of interval (stretch, *215) and of derivatives (*216). These two first sections contain the tools that will be used in the remaining parts. Section C deals with the foundations of real analysis: Russell and Whitehead introduce convergence (*230), limit (*233) and continuity of a function (*234). Sections D and E are devoted to Cantor's theory of ordinals, that is, to the study of the properties of well-ordered series (recall that Russell and Whitehead developed in PM IV a relation-arithmetic, more general than Cantor's ordinal arithmetic). Section F deals with compact series. The three last sections are then concerned with two species of series: the well-ordered ones and the compact ones.

20. See also (1912, p. 515): 'section C ... is concerned with convergence and the limits of functions and the definition of a continuous function. Its purpose is to show how these notions can be expressed, and many of their properties established, in a much more general way than is usually done, and without assuming that the arguments or values of the functions concerned are either numerical or numerically measurable.' One finds exactly the same idea in *Introduction to Mathematical Philosophy*. At the beginning of Chapter 11, Russell explained (1919, p. 107): 'limits and continuity of functions, in works on ordinary mathematics, are defined in terms involving number. This is not essential, as Dr Whitehead has shown [see *Principia Mathematica*, vol. ii, *230–4]. We will, however, begin with the definitions in the text-books, and proceed afterwards to show how these definitions can be generalized so as to apply to series in general, and not only to such as are numerical or numerically measurable.'

21. Let me quote the introduction of Hausdorff (1906, p. 45): 'as far as I know, all the *essential* results of these works are new since the hitherto existing investigations of others refer almost exclusively to subsets of the linear continuum or to well-ordered sets ... With that said, it ought not be disputed that details from the following investigations may already be found in the existing literature ... However, the correspondence of my abstract investigations with those of others, whose main purpose is the application to analysis and geometry, probably does not extend beyond such analysis.' For a very interesting analysis of Cantor's ordinal theory and its reception, see Hallett (1984).

22. Let me quote this passage of a letter from Baire to Borel (31 January 1908): 'Molk [the editor of Baire (1909)] must have been disappointed by certain parts of my paper, as for instance the one which deals with Hausdorff's work, in which Alephs spread across the horizon' (Baire 1990, p. 88).

23. Russell and Whitehead refer to Hausdorff in PM. They share with him the general perspective according to which the theory of order types should be developed independently of the axiom of choice and independently of any consideration about the application (see the correspondence with Jourdain in Grattan-Guinness 1977). But they also borrow from Hausdorff some techniques (like the 'principle of first difference' (PM, 1912, pp. 403–7)) and some results (on compact series (PoM, p. 171)).

24. Russell and Whitehead elaborated a general theory of relation type in PM IV. The impetus was that they did not wholeheartedly accept the axiom of choice and did not then consider that any set can be well-ordered; they thus considered

that a more general theory of order type was needed. That's why they elaborated a theory of relation numbers, which is presented as a generalization of Cantor's theory of ordinals.

25. In 1903, Russell's account of real analysis is based on Dini's *Fundamenti per la teorica delle functioni delle variabili reali* (1878). In chapter 39 of PoM, Russell resumed the definitions of the continuity of function (§ 304), the differential coefficient (§ 305) and the integral (§ 307). In PM, Dini's definition of the continuity of a function is maintained (see *234), but Russell and Whitehead make it clear that Dini's definition can 'be so stated as to be free from any reference to number, and derivable from the ideas dealt with in the previous numbers of the present section'.

26. One finds in Wiener (1921) a PM inspired theory of measurement, where the result of a measurement is seen as a construction from the rough data of experience. On this, see Gandon (2009e).

27. I have explained above that, in 1903, Russell gave an argument explaining what makes the progression made up of finite cardinals so special: if Peano's axioms suffice to develop arithmetic, they do not account for the use of whole numbers in counting. As we saw, no analogous argument was developed to justify the introduction of real numbers in PoM, chapter 33. In PM, the reference to measurement played the role that the reference to counting filled in PoM: it gave the study of real numbers its *raison d'être*.

28. Russell also based his definition of whole numbers on *Applic*. And the use of *Applic* in this case has many points in common with its use for justifying the relational definition of real in PM. Let me quote PoM (p. 251): 'thus it is plain that ordinals, either as classes of like serial relations, or as notions like "nth", are more complex than cardinals; that the logical theory of cardinals is wholly independent of the general theory of progressions, requiring independent development in order to show that the cardinals form a progression; and that Dedekind's ordinals are not essentially either ordinals or cardinals, but the members of any progression whatever'. *Applic* can thus be seen as a means to establish that the 'logical theory of cardinals is wholly independent of the general theory of progressions': it would be because one can apply the concept of '2' to a class, without referring to the notion of '2nd', that the theory of cardinal should be distinguished from the theory of ordinal. This would parallel the distinction between real number and continuous series in the real number case.

29. Abstraction principles are second-order formulae of the form:

$$\forall F \forall G(\S(F) = \S(G) \Leftrightarrow F \text{ eq } G),$$

where §( ) is some operator which, when applied to a predicate variable, forms a term, and eq is short for some definable equivalence relation between F and G (see for example the presentation of Hale's construction on pp. 139–41).

30. The literature about neo-logicism is immense. For a useful overview of the questions raised by this programme see Hale and Wright (2005). For good introductions, see Heck (1998) and Boolos (1998).

31. See Shapiro (2000, p. 336): 'my purpose is to present other (logical) abstraction principles that can be employed to develop a theory of the real numbers in much the same way that Hume's Principle yields a theory of the natural numbers. The crucial aspect of the treatment – where terminology for real numbers is introduced – roughly follows the development in Dedekind's celebrated *Stetigkeit und irrational Zahlen*, but I formulate the relevant existence principles as Fregean abstractions rather than Dedekind-type structuralist principles.'

32. For more detail, see Wright (2000).

33. It is important to realize that Shapiro shares with the neo-logicists the idea that, from a philosophical point of view, satisfying the epistemological task is not enough – that the metaphysical requirement should also be met. This is what distinguishes his 'metaphysical' brand of structuralism from some other versions – like for instance the one endorsed by Benacerraf (1965).

34. Shapiro traces back this demand to Dedekind, who refused to identify reals with cuts, because cuts are sets and have members, and that it is odd for a mathematician to assume that real numbers have members. See Shapiro (2006, p. 112): 'Dedekind [said] that there are many properties that cuts have which would sound very odd if applied to the corresponding real numbers ... For example, cuts have members. Do real numbers have members? Dedekind's Benacerraf-type point makes sense in the context of the minimalism constraint ... In the Preface to his later treatise on the natural numbers, Dedekind (1888) agreed that Cantor and Weierstrass both gave "perfectly rigorous", and so presumably acceptable, accounts of the real numbers, even though they were different from his own and from each other. But Dedekind wrote that his approach is "somewhat simpler, I might say quieter" than the others. I presume he meant that his own account of the real numbers has less excess baggage than the others. This notion of "quietness" is consonant with the minimalism constraint.'

35. See Shapiro (2006) for more about this.

36. See Wright (2000, p. 329): '[there is] no significant shortcoming, from the neo-Fregean point of view, in an abstractionist reconstruction of the reals that follows the Dedekindian way'.

37. See Wright (2000, p. 320): 'the fact is that Hume's Principle accomplishes two quite separate tasks. There is, *a priori*, no particular reason why a principle intended to incorporate an account of the nature of a particular kind of mathematical entity should also provide a sufficient axiomatic basis for the standard mathematical theory of that kind of entity ... [The] striking feature of the neo-Fregean foundations for number theory is that the one core principle, Hume's Principle, discharges *both* roles. This is not a feature which we should expect to be replicated in general when it comes to providing abstractionist foundations for other classical mathematical theories.'

38. The fact that numerals function as singular terms is by itself a sufficient reason to consider them as singular terms and then as mathematical objects. See Hale and Wright (2001, p. 8) and also the first two chapters of Hale (1988).

39. This insight has been acknowledged by many as one of the main contributions Russell made to twentieth-century philosophy. See for instance Wittgenstein (1921, 4.0031): 'Russell's merit is to have shown that the apparent logical form of the proposition need not be its real form'. See also Ramsey (1990) and Quine (1953). Hylton (1990) has put this insight at the centre of his reconstruction.

40. This development does not pretend to make sense of the role of the acquaintance principle in Russell. The only point I want to make is that, in PoM, the real form is reached through an analysis of the sentence, and not by a direct non-linguistic road. For a discussion of the role of the acquaintance principle, see Wahl (2007).

41. For an introduction to Plücker's works, see Brieskorn and Knörrer (1986).

7 Russell's Universalism and Topic-Specificity

1. This is the area surrounding the issue of the criteria of identity of *Sinn* in Frege's scholarship. For a lucid commentary, see Blanchette (2012).

2. Weiner (1990 and 2007) has heavily relied on this 1914 passage in her interpretation of Frege's philosophy. For a powerful criticism of Weiner, see Blanchette (2012). It is not clear to me how much weight one should put on Frege (1914), and I certainly do not want to extract from what precedes a global interpretation of his work. As I have said, I am using Frege to understand Russell better – I do not attempt to develop an interpretation of Frege.

3. There are of course various interpretations of Wittgenstein's remarks on decimal notation. For more on this, see Marion (1998) and also Mühlhölzer (2005). As for Frege, I take Wittgenstein here as a spokesman of a position that he perhaps never endorsed.

4. See as well Goldfarb (1989) and Hylton (1990).

5. I thank Tom Jones for the translations of Rivenc's texts.

6. In particular, Rivenc is very close to van Heijenoort's construal of universalism – the way he presented universalism should then be refined in the light of more recent developments.

7. Russell's theory of number and quantity in PM VI illustrates this in a very clear way. Recall that Russell justified his definition of ratio as relations in section A by referring to the theory of quantitative measurement, examined in section C. The right point of view is then neither the one one has at the end of the process (in section C), nor the one one has at the beginning of the construction (in section A). To understand PM VI is precisely to realize that what appears in section A as the result of an extra-systematic demand, as a 'useless appendage', is at a later stage internalized within the system. At the same time, the reader of section C should not forget the perspective he or she had in section A: *Applic* is what justifies the whole construction.

8. Russell and Whitehead underlined in the introduction of PM (1910, p. vi) this back and forth movement between existing mathematics and the logical system: 'in constructing a deductive system such as that contained in the present work, there are two opposite tasks which have to be concurrently performed. On the one hand, we have to analyse existing mathematics, with a view to discovering what premisses are employed, whether these premisses are mutually consistent, and whether they are capable of reduction to more fundamental premisses. On the other hand, when we have decided upon our premisses, we have to build up again as much as may seem necessary of the data previously analysed, and as many other consequences of our premisses as are of sufficient general interest to deserve statement'. See as well PoM (p. xx) and Russell (1919, ch. 1). The importance of Russell's regressive method has been underlined by Irvine (1989). For more on Irvine's interpretation, see note 12.

9. See also p. 2: 'empires – self-consciously maintaining the diversity of people they conquered and incorporated – have played a long and critical part in human history ... Empires, of course, hardly represented a spontaneous embrace of diversity. Violence and day-to-day coercion were fundamental to how empires were built and how they operated. But as successful empires turned their conquests into profit, they had to manage their unlike populations, in the process producing a variety of ways to both exploit and rule.'

10. Rawls (1971, pp. 18–20, 40–7) is responsible for the reintroduction of this notion in contemporary philosophy. See for instance (p. 19): 'a conception of justice cannot be deduced from self-evident premises or conditions on principles; instead, its justification is a matter of the mutual support of many considerations, of everything fitting together into one coherent view'. What is said about our conception of justice holds for Russell's analysis of mathematics. It is interesting to note that Rawls refers to Sidgwick as a forefather of his method. Now, as is

well known, Russell and Moore followed Sidgwick's classes. And it seems to me that Sidgwick's influence on their view of the philosophical method has not been taken sufficiently into account. Moore (1903) is not in disagreement with Sidgwick's methodology, even if he has some reservations about some of his conclusions; see especially (1903, § 45).

11. As I have said, my aim is not to give an interpretation of Frege's work. And in certain respects, the reading I have expounded oversimplifies Frege's position. Thus, the whole point of Frege (1884) is to convince his readers, by very different sorts of argument, of the acceptability of his logical translation (on this, see Blanchette 2012). The difference between Russell and the historical Frege is thus certainly smaller than what I have suggested here. But there is still a difference. First, the fact that Frege attempted to persuade his readers of the correctness of his analysis in a separate book (not in the presentation of his system) could be interpreted as a way of emphasizing that this task does not pertain to the logicist project, strictly considered. This would be in line with Frege (1914). Secondly and perhaps more importantly, Frege was first and foremost interested in reducing arithmetic to logic. He did not believe that geometry was a part of logic. The mathematics he considered was then much more homogeneous than the mathematics Russell attempted to analyse. As a consequence, the architectonic question (how to structure the mathematical material) was not, for Frege, as pressing as it was for Russell. This difference of target could explain the difference of strategy.

12. I would like to add a final word on the relation between my interpretation and the one defended by Irvine (1989). The two seem at first very close, in so far as they both insist on the circularity of Russell's reasoning. Irvine thus explained that, in PoM and PM, the logical principles are epistemologically justified by the theorems which are logically deduced from them (pp. 311–12). He quotes Russell (1907b, p. 273 ff.): 'we tend to believe the premises because we can see that their consequences are true, instead of believing the consequences because we know the premises to be true. But the inferring of premises from consequences is the essence of induction; thus the method in investigating the principles of mathematics is really an inductive method, and is substantially the same as the method of discovering general laws in any other science.' If this emphasis on the circularity of Russell's justification process is in line with what I have defended here, there is an important difference between the two readings: Irvine focuses on truth and justification; I focus on meaning and 'content'. My aim was to establish that the content of a mathematical proposition (not its justification) is neither completely given in the pre-logicized mathematics, nor provided by logic. This difference is an important one, since the focus on justification tends to divert attention from the issue concerning the identification of content. One usually does not inquire about the grounds of a proposition, when its subject matter is not fixed. Irvine's interpretation could then lead us to believe that the logicization process does not affect the content of the pre-logicized mathematics. As I have argued in this book, this was however a central issue for Russell.

13. Far from being critical of the arithmetical definition of the real numbers, Suppes's theory of measurement seems to presuppose it. For more on this, see pp. 155–6, 228, 235.

14. These questions are partly analysed in the literature; see for instance Landini (1998) and Linsky (1999).

15. See, among others, Corfield (2003), Ferreirós and Gray (2006), Grosholz (2007), Mancosu (2008) and Van Kerkhove (2009). Mancosu traces the distinction

between 'mainstream' foundationalism and practice-based philosophy back to the works of Lakatos (and then Kitcher, Tymoczko, etc.) in the 1960s. The later Wittgenstein could also be regarded as a forefather of this attempt to return to the 'rough ground' of real mathematics.

16. The formula is from Corfield (2003).

17. See Wittgenstein (1956) and Ferreirós (1999), Ferreirós and Gray (2006) and Grosholz (2007). Note however that Corfield (2003, pp. 236, 269–70) does not share this view.

18. For more on the cognitive theory of numbers, see Dehaene (1999); for more on the cognitive approach of diagram and visualization, see the contributions in Mancosu et al. (2005).

19. For more on this line of research, see Buldt et al. (2008) and Löwe and Müller (2008).

20. For a criticism of the practice-based approaches that are close to what I am saying here, see Paseau (2005), Arana (2007) and Larvor (2010).

21. See for instance Mancosu (2008, p. 26), who distances himself in this respect from Corfield: 'what is distinctive of this volume is that we integrate local studies with general philosophy of mathematics'.

22. See, among others, Diamond (1995).

23. One can, for instance, consider that, in the early modern period, the theory of the powers of the mind represented a framework which enabled philosophers and mathematicians to develop a general conception of mathematics in tune with the mathematical researches of the time. The distinction between reason, imagination and sensibility allowed philosophers to encroach their thought onto the 'real' mathematics without losing sight of the global articulation between the different mathematical domains. This setting was not logically centred – but it was general and flexible enough to play the role that the logical framework played in Russell. For an analysis of Kant's philosophy of mathematics along this line, see Friedman (1992).

24. Rayo (2005) is a perfect illustration of this standard approach. Logicism is here presented as a family of theorems, which hold or do not hold according to the background logic one adopts.

25. A more balanced way to draw this contrast would be to say that Frege's logicism was centred on proofs, while Russell's logicism was centred on definitions. For an insightful discussion of Frege's notion of proof, see Blanchette (2012, ch. 1).

Bibliography

Alvarez C. and Arana A. (eds). Forthcoming. *Analytic Philosophy and the Foundations of Mathematics*, Basingstoke, Palgrave Macmillan.

Anderson C. A. and Savage C. W. 1989. *Rereading Russell: Essays in Bertrand Russell's Metaphysics and Epistemology*, Minneapolis, University of Minnesota Press.

Antonelli A. and May R. 2000. Frege's New Science, *Notre Dame Journal of Formal Logic*, 41(3): 242–70.

Arana A. 2007. Review of D. Corfield's *Towards a Philosophy of Real Mathematics*, *Mathematical Intelligencer*, 29(2): 80–3.

Armstrong D. 1978. *Universals and Scientific Realism*, Cambridge, Cambridge University Press.

Armstrong D. 1988. Are Quantities Relations? A Reply to Bigelow and Pargetter, *Philosophical Studies*, 54: 305–16.

Arnauld A. and Nicole P. 1683. *La logique ou l'art de penser*, ed. and tr. in English by J. V. Buroker, Cambridge University Press, 1996.

Artin E. 1958. *Geometric Algebra*, New York, Interscience Publisher.

Artin E. and Schreier O. 1926. Algebraische Konstruction reeller Körper, *Abhandlungen aus dem Mathematischen Seminar der Universität Hamburg*, 5: 85–99.

Aspray W. and Kitcher P. eds 1988. *History and Philosophy of Modern Mathematics*, Minneapolis, Minnesota Studies in the Philosophy of Science, vol. XI.

Avellone M., Brigaglia A. and Zapulla C. 2002. The Foundations of Projective Geometry in Italy from De Paolis to Pieri. *Archive for History of Exact Sciences*, 56: 363–425.

Avigad J. 2008. Understanding Proofs, in Mancosu (2008), pp. 317–53.

Baire R. 1990. Lettres de René Baire à Emile Borel, *Cahiers du séminaire d'histoire des mathématiques*, vol. 11, pp. 33–120.

Baire R. and Schoenflies A. 1909. Théorie des ensembles, in Molk (1904–16), vol. I-1, pp. 489–531.

Barrow-Green J. and Gray J. 2006. Geometry at Cambridge 1863–1940, *Historia Mathematica*, 33: 315–56.

Batitsky V. 1998. Empiricism and the Myth of Fundamental Measurement, *Synthese*, 116: 51–73.

Batitsky V. 2002. Some Measurement-Theoretic Concerns about Hale's 'Reals by Abstraction', *Philosophia Mathematica*, 10: 286–303.

Benacerraf P. 1965. What Numbers Could Not Be, *Philosophical Review*, 74: 47–73.

Benacerraf P. 1973. Mathematical Truth, *Journal of Philosophy*, 70: 661–79.

Benacerraf P. 1981. Frege: The Last Logicist, *Midwest Studies in Philosophy*, 6(1): 17–36.

Bettazzi R. 1890. *Teoria delle Grandezze*, Pisa, Enrico Spoerri.

Bigelow J. 1988. *The Reality of Numbers. A Physicalist's Philosophy of Mathematics*, Oxford, Oxford University Press.

Bigelow J. and Pargetter R. 1988. Quantities, *Philosophical Studies*, 54(3): 287–304.

Blanchette P. 1996. Frege and Hilbert on Consistency, *Journal of Philosophy*, XCIII(7): 317–36.

Blanchette P. 2007. Frege on Consistency and Conceptual Analysis, *Philosophia Mathematica*, 15(3): 321–46.

Blanchette P. 2012. *Frege's Conception of Logic*, Oxford, Oxford University Press.

Boolos G. 1986. Saving Frege From Contradiction, *Proceedings of the Aristotelian Society*, 87, pp. 137–51.

Boolos G. 1990. The Standard of Equality of Numbers, in G. Boolos ed., *Meaning and Method: Essays in Honor of Hilary Putnam*, Cambridge, Cambridge University Press, pp. 261–77.

Boolos G. 1994. The Advantages of Honest Toil over Theft, in A. George, ed., *Mathematics and Mind*, Oxford, Oxford University Press, pp. 27–44.

Boolos G. 1998. Gottlob Frege and the Foundations of Arithmetic, in J. Burgess ed., *Logic, Logic and Logic*, Cambridge (MA), Harvard University Press, pp. 143–54.

Bosanquet B. 1888. *Logic or the Morphology of Knowledge*, Oxford, Clarendon Press.

Bourbaki N. 1950. The Architecture of Mathematics, *American Mathematical Monthly*, 57(4): pp. 221–32.

Boutroux Pierre 1905. Correspondance mathématique et relation logique, *Revue de Métaphysique et de Morale*, 13: 620–37.

Boutroux Pierre 1914–1919. *Les principes de l'analyse mathématique. Exposé historique et critique*, 2 volumes, Paris, Hermann.

Brannan A., Esplen M. F. and Gray J. 1999. *Geometry*, Cambridge, Cambridge University Press.

Bressoud D. M. 2008. *A Radical Approach to Lebesgue's Theory of Integration*, Cambridge, Cambridge University Press.

Brieskorn E. and Knörrer H. 1986. *Plane Algebraic Curves*, tr. John Stillwell, Basel, Birkhäuser.

Brunschvicg L. 1912. *Les étapes de la philosophie des mathématiques*, Paris, Alcan, 1912.

Buekenhout F. 1995. *Handbook of Incidence Geometry*, Amsterdam, Elsevier Science.

Buldt B., Löwe B. and Müller T. 2008. Towards a New Epistemology of Mathematics, *Erkenntnis*, 68: 309–29.

Burali-Forti C. 1893. Sulla teoria delle grandezze, *Rivista di Matematica*, 3: 76–101.

Burali-Forti C. 1897. *Introduction à la géométrie différentielle suivant la méthode de H. Grassmann*, Paris, Gauthier-Villars.

Burali-Forti C. 1898. Les propriétés formales des opérations algébriques, *Rivista di Matematica*, 6: 141–77.

Burali-Forti C. 1903. Sulla teoria general delle grandezze e dei numeri, *Atti della R. Accademia delle Scienze di Torino*, 39: 256–72.

Burbank J. and Cooper F. 2010. *Empires in World History. Power and the Politics of Difference*, Princeton, Princeton University Press.

Byrd M. 1985. Part I of *The Principles of Mathematics*, *Russell: The Journal of Bertrand Russell Studies*, 4, pp. 271–88.

Byrd M. 1987. Part II of *The Principles of Mathematics*, *Russell: The Journal of Bertrand Russell Studies*, 7, pp. 60–70.

Byrd M. 1994. Part V of *The Principles of Mathematics*, *Russell: The Journal of Bertrand Russell Studies*, 14, pp. 47–86.

Byrd M. 1996. Part III–IV of *The Principles of Mathematics*, *Russell: The Journal of Bertrand Russell Studies*, 16, pp. 145–68.

Cantor G. 1895–97. Beiträge zur Begründung der transfiniten Mengenlehre, *Mathematische Annalen*, 46: 481–512; 49: 207–46.

Carus S. and Fano G. 1915. *Exposé parallèle du développement de la géométrie synthétique et de la géométrie analytique pendant le XIXe siècle*, in Molk (1904–16), vol. III-1, pp. 185–259.

Carnot L. 1804. *Géométrie de position*, Paris, Duprat.

Cassirer E. 1910. *Substanzbegriff und Funktionsbegriff: Untersuchungen über die Grundfragen der Erkenntniskritik*. Berlin, Bruno Cassirer. Translated as *Substance and Function*. Chicago, Open Court, 1923.

Cayley A. 1859. Sixth Memoir upon Quantics, *Philosophical Transactions of the Royal Society of London*, 149: 61–90.

Cayley A. 1878. On the Theory of Groups, *Proceedings of the London Mathematical Society*, 9: 126–33.

Chang H. 2004. *Inventing Temperature: Measurement and Scientific Progress*, Oxford, Oxford University Press.

Coffa A. 1981. Russell and Kant, *Synthese*, 46: 247–63.

Corfield D. 2003. *Towards a Philosophy of Real Mathematics*, Cambridge, Cambridge University Press.

Corry L. 1989. Linearity and Reflexivity in the Growth of Mathematical Knowledge, *Science in Context*, 3: 409–40.

Corry L. 1996. *Modern Algebra and the Rise of Mathematical Structures*, Basel and Boston, Birkhäuser Verlag.

Couturat L. 1896. *De l'Infini mathématique*, Paris, Alcan.

Couturat L. 1900. Études critiques. L'algèbre universelle de M. Whitehead, *Revue de métaphysique et de morale*, 8(1900): 323–62.

Couturat L. 1905. *Les Principes des Mathématiques*, Paris, Alcan.

Coxeter H. S. M. 1947. *Non-Euclidean Geometry*, 2nd edn, Toronto, Toronto University Press.

Coxeter H. S. M. 1949. *The Real Projective Plane*, York, Maple Press Company.

Cremona L. 1885. *Elements of Projective Geometry*, tr. Charles Leudesdorf, Oxford, Oxford University Press.

Crowe M. J. 1967. *A History of Vector Analysis: The Evolution of the Idea of a Vectorial System*, Notre Dame, Notre Dame University Press.

Darrigol O. 2003. Number and Measure: Hermann von Helmholtz at the Crossroads of Mathematics, Physics and Psychology, *Studies in History and Philosophy of Science*, 34: 515–73.

Dedekind R. 1872. Continuity and Irrational Numbers, in Dedekind (1963), pp. 1–30.

Dedekind R. 1888. The Nature and Meaning of Numbers, in Dedekind (1963), pp. 31–106.

Dedekind R. 1963. *Essay on the Theory of Numbers*, Mineola, Dover.

Dehaene S. 1999. *The Number Sense: How the Mind Creates Mathematics*, Oxford, Oxford University Press.

Dembowski P. 1968. *Finite Geometries*, Berlin, Springer.

Demopoulos W. 1994. Frege, Hilbert, and the Conceptual Structure of Model Theory, *History and Philosophy of Logic*, 15(2): 211–25.

Demopoulos W. 1995. *Frege's Philosophy of Mathematics*, Cambridge (MA), Harvard University Press.

Desmet R. 2010. Whitehead's Philosophy of Mathematics and Relativity, PhD Dissertation, Ghent University.

Detlefsen M. 1993. Poincaré vs. Russell on the role of logic in mathematics, *Philosophia Mathematica*, 1(1): 24–49.

Detlefsen M. 2007. Purity as an Ideal of Proof, in Mancosu (2008), pp. 248–65.

Diamond C. 1995. *The Realistic Spirit: Wittgenstein, Philosophy and the Mind*, Palatino, Bradford Books.

Diez J. A. 1997. A Hundred Years of Numbers, An Historical Introduction to Measurement Theory 1887–1990, *Studies in History and Philosophy of Science*, 28: 167–85, 237–65.

Dreben B. and van Heijenoort J. 1986. Introductory notes in K. Gödel *Collected Works: Volume I (Publications 1929–1936)*, Feferman S. *et al.* eds., Oxford, Oxford University Press.

Dini U. 1878. *Fondamenti per la teorica delle functioni delle variabili reali*, Pisa, Nistri & Cie.

Dugac P. 2003. *Histoire de l'analyse: autour de la notion de limite et de ses voisinages*, Paris, Vuibert.

Dummett M. 1973. *Frege: Philosophy of Language*, Cambridge (MA), Harvard University Press.

Dummett M. 1991. *Frege: Philosophy of Mathematics*, Cambridge (MA), Harvard University Press.

Dummett M. 1995. *Frege's Theory of Real Numbers*, repr. of ch. 22 of Dummett (1991) in Demopoulos (1995).

Ehrlich P. 2006. The Rise of Non-Archimedean Mathematics and the Roots of a Misconception I: The Emergence of Non-Archimedean Systems of Magnitudes, *Archive for History of Exact Sciences*, 60, pp. 1–121.

Epples M. 2003. The End of the Science of Quantity: Foundations of Analysis (1860–1910), in H. N. Jahnke ed., *A History of Analysis*, Providence, AMS Bookstore, pp. 291–324.

Ewald W. B. ed. 1996. *From Kant to Hilbert: A Source Book in the Foundations of Mathematics*, Oxford, Oxford University Press.

Ferreirós J. 1999. *Labyrinth of Thought. A History of Set Theory and its Role in Modern Mathematics*, Basel, Birkhäuser.

Ferreirós J. and Gray J. J. 2006. *The Architecture of Modern Mathematics*. Oxford, Oxford University Press.

Field H. 1980. *Science Without Numbers: A Defence of Nominalism*, Princeton, Princeton University Press.

Forster M. B. 1931. The Concrete Universal: Cook Wilson and Bosanquet, *Mind*, 40(157): 1–22.

Frege, G. 1879. *Begriffsschrift: eine der arithmetischen nachgebildete Formelsprache des reinen Denkens*, Halle, L. Nebert (quoted from T. W. Bynum tr. and ed., Oxford, Oxford University Press, 1972).

Frege, G. 1884. *Grundlagen der Arithmetik*, Breslau, W. Koebner (quoted from J. L. Austin tr. and ed., Oxford, Basil Blackwell, 1974).

Frege, G. 1891. Funktion und Begriff, Address given to the *Jenaische Gesellschaft für Medecin und Naturwissenschaft* (quoted from the English tr. in Geach (1980), pp. 21–41).

Frege, G. 1892. Über Sinn und Bedeutung, *Zeitschrift für Philosophie und Philosophische Kritike*, 100 (quoted from the English tr. in Geach (1980), pp. 56–78).

Frege, G. 1893. *Grundgesetze der Arithmetik*, vol. 1, Jena, Hermann Pohle, repr. Hildesheim, Olms, 1966.

Frege, G. 1903a. Über die Grundlagen der Geometrie, *Jahrbericht der deutschen Mathematiker Vereinigung*, 12 (tr. in Frege (1971), pp. 22–37).

Frege, G. 1903b. *Grundgesetze der Arithmetik*, vol. 2, Jena, Hermann Pohle repr. Hildesheim, Olms 1966.

Frege, G. 1914. Logic in Mathematics, MS published in Frege (1979), pp. 203–50.

Frege, G. 1971. *On the Foundations of Geometry and Formal Theories of Arithmetic*, tr. and ed. E.-H. Kluge, Yale, Yale University Press.

Frege, G. 1979. *Posthumous Writings*, ed. Hermes *et al.*, Chicago, University of Chicago Press.

Frege, G. 1980. *Philosophical and Mathematical Correspondence of Gottlob Frege*, ed. B. McGuinness, Chicago, University of Chicago Press.

Friedman M. 1992. *Kant and the Exact Sciences*, London, Harvard University Press.

Gandon S. 2004. Russell et l'*Universal Algebra* de Whitehead: la géométrie projective entre ordre et incidence (1898–1903), *Revue d'histoire des mathématiques*, 10: 187–256.

Gandon S. 2005a. Algèbre, géométrie et loi d'intensité: l'enjeu de *A Treatise on Universal Algebra*, in Michel Weber and Diane d'Eprémesnil eds, *Chromatikon I, Annuaire de la philosophie en procès – Yearbook of Philosophy in Process*, Presses Universitaires de Louvain.

Gandon S. 2005b. Pasch entre Klein et Peano: empirisme et idéalité en géométrie, *Dialogue*, 14: 653–92.

Gandon S. 2006a. Grandeurs, vecteurs et relations chez Russell (1897–1903), *Philosophiques*, 33–2: 333–62.

Gandon S. 2006b. La reception des *Vorlesungen über neuere Geometrie* de Pasch par Peano, *Revue d'histoire des mathématiques*, 12(2006): 249–90.

Gandon S. 2008. Which Arithmeticisation for which Logicism? Russell on Quantities and Relations, *History and Philosophy of Logic*, 29(1): 1–30.

Gandon S. 2009a. Toward a Topic-Specific Logicism? Russell's Theory of Geometry in the *Principles of Mathematics*, *Philosophia Mathematica*, 17(1): 35–72.

Gandon S. 2009b. Russell, les '*sense data*' et les objets physiques: une approche géométrique de la classification, *Philosophae scientiae*, 13(1): 71–97.

Gandon S. 2009c. Wittgenstein dans la fabrique des *Principia*: sur et autour *Tractatus* 3. 33, C. Chauviré éd., *Lire le Tractatus-Logico-Philosophicus de Wittgenstein*, Vrin, 2009, 91–120.

Gandon S. 2009d. La théorie des rapports chez Augustus de Morgan, *Revue d'histoire des sciences*, 62(1): 285–312.

Gandon S. 2009e. Relation et quantité chez Russell (1897–1913), unpublished MS written for the 'habilitation à diriger des recherches', Clermont University.

Gandon S. 2010a. Peano's Logical Language and Grassmann's Legacy, in C Roero *et al.* eds, *Actes du Congresso Internazionale Giuseppe Peano e la sua Scuola, fra matematica, logica e interlingua*, Torino, DSSP.

Gandon S. 2010b. Entre figure et espace: le cas des diagrammes en géométrie finie, online website 'images des mathématiques', http://images.math.cnrs.fr/Entrefigures-et-espaces-le-cas.html

Gandon S. Forthcoming. Logicism and Mathematical Practices – Russell's Theory of Metrical Geometry in *The Principles of Mathematics*, in C. Alvarez and A. Arana (forthcoming).

Geach P. 1980. *Translations from the Philosophical Writings of Gottlob Frege*, 3rd edn, P. Geach and M. Black eds and trs, Oxford, Blackwell.

Godwyn M. and Irvine A. D. 2003. Bertrand Russell's Logicism, in Griffin (2003), pp. 171–202.

Goldfarb W. 1979. Logic in the Twenties: The Nature of the Quantifier, *Journal of Symbolic Logic*, 44: 351–68.

Goldfarb W. 1989. Russell's Reasons for Ramification, in Anderson and Savage (1989), pp. 24–40.

Grabiner J. 1981. *The Origins of Cauchy's Rigorous Calculus*, Cambridge (MA), MIT Press.

Grassmann H. 1844. *Die Lineale Ausdehnungslehre, eine neuer Zweig der Mathematik, dargestellt und durch Anwendungen auf die übrigen Zweige der Mathematik, wie auch auf die*

Statik, Mechanik, die Lehre vom Magnetismus und die Kristallonomie erläutert, Leipzig, Verlag von Otto Wigand (French tr. by D. Flament, Paris, Blanchard, 1994).

Grassmann H. 1861. *Lehrbuch der Arithmetik für höhere Lehranstalten*, Berlin, Enslin.

Grassmann H. 1862. *Die Ausdehnungslehre. Vollständig und in strender Form bearbeitet*, Berlin, Enslin (English tr. by L. C. Kannenberg, Providence, American Mathematical Society, 2000).

Grattan-Guinness I. 1977. *Dear Russell – dear Jourdain. A commentary on Russell's logic, based on his correspondence with Philip Jourdain*, New York, Columbia University Press.

Grattan-Guinness I. 2000. *The Search for Mathematical Roots, 1870–1940*, Princeton, Princeton University Press.

Griffin N. 1985. Russell's Multiple Relation Theory of Judgment. *Philosophical Studies*, 47(2): 213–47.

Griffin N. 1991. *Russell's Idealist Apprenticeship*, Oxford, Clarendon Press.

Griffin N. 2002. Russell, Logicism, and 'If-thenism', in A. Schwerin ed., *Bertrand Russell on Nuclear War, Peace, and Language*, Westport, Praeger.

Griffin N. 2003a. Russell's Philosophical Background, in Griffin (2003), pp. 84–107.

Griffin N. ed. 2003b. *The Cambridge Companion to Bertrand Russell*, Cambridge, Cambridge University Press.

Grosholz E. 1981. Wittgenstein and the Correlation of Logic and Arithmetic, *Ratio*, 23(1): 31–42.

Grosholz E. 2007. *Representation and Productive Ambiguity in Mathematics and the Sciences*, Oxford, Oxford University Press.

Guigon G. 2006. Meinong on Magnitudes and Measurement, *Meinong Studies*, V(1): 255–96.

Hadamard J. 1945. *Psychology of Invention in the Mathematical Field*, Princeton, Princeton University Press.

Hager P. 1994. *Continuity and Change in the Development of Russell's Philosophy*, Boston and London, Kluwer Academic Publishers.

Hahn H. 1907. Über die nichtarchimedischen Grössensysteme, *Sitzungsberichte der Kaiserlichen Akademie der Wissenschaften, Wien, Mathematisch – Naturwissenschaftliche Klasse*, 116 (Abteilung IIa), pp. 601–55.

Hale B. 1988. *Abstract Objects (Philosophical Theory)*, London, Blackwell.

Hale B. 1996. Structuralism's Unpaid Epistemological Debts, *Philosophia Mathematica*, 4: 124–47.

Hale B. 2000. Reals by Abstraction, *Philosophia Mathematica*, 8: 100–23.

Hale B. 2002. Real Numbers, Quantities, and Measurement, *Philosophia Mathematica*, 10: 304–23.

Hale B. and Wright C. 2001. *The Reason's Proper Study: Essays Toward a Neo-Fregean Philosophy of Mathematics*, Oxford, Clarendon Press.

Hale B. and Wright C. 2002. Benacerraf's Dilemma Revisited, *European Journal of Philosophy*, 10(1): 101–29.

Hale B. and Wright C. 2005. Logicism in the Twentieth-First Century, in S. Shapiro ed., *The Oxford Handbook for Logic and the Philosophy of Mathematics*, Oxford, Oxford University Press, pp. 166–202.

Hallett M. 1984. *Cantorian Set Theory and Limitation of Size*, Oxford, Clarendon Press.

Hallett M. and Majer U. 2004. *David Hilbert Lectures on the Foundations of Geometry 1891–1902*, Heidelberg and New York, Springer-Verlag.

Hardy G. H. 1910. *Orders of Infinity, The 'Infinitärcalcül' of Paul Du Bois-Reymond*, Cambridge, Cambridge University Press.

Hardy G. H. 1949. *Divergent Series*, Oxford, Oxford University Press.

Harrell M. 1988. Extension to Geometry of *Principia Mathematica* and Related Systems II, *Russell: The Journal of Bertrand Russell Studies*, 8, pp. 140–60.

Hausdorff F. 1906. Investigations into Order Types I, II, III, in Plotkin (2005), pp. 45–95.

Hawkins T. 1970. *Lebesgue Theory of Integration*, Madison, University of Wisconsin Press.

Heck R. G. 1993. The Development of Arithmetic in Frege's *Grundgesetze der Arithmetik*, *Journal of Symbolic Logic*, 58, pp. 579–601.

Heck R. G. 1997. Finitude and Hume's Principle, *Journal of Philosophical Logic*, vol. 26, pp. 589–617.

Heck R. G. 1998. Introduction au théorème de Frege, in M. Marion and A. Voizard, eds, *Frege: Logique et philosophie*, Harmattan, pp. 33–61.

Heidelberger M. 2004. *Nature From Within: Gustav Theodor Fechner and His Psychophysical Worldview*. Pittsburgh, University of Pittsburgh Press (Tr. from the German by C. Klohr).

Helmholtz H. L. 1887. Zählen und Messen, erkenntnisstheoretisch betrachtet, *Philosophische Aufsätze, Eduard Zeller zu seinem fünfzigjährigen Doctorjubiläum gewidmet*, Leipzig, Fues (Quoted from, Helmholtz (1977), pp. 72–102).

Helmholtz H. L. 1977. *Hermann von Helmholtz: Epistemological Writings. The Paul Hertz/ Moritz Schlick Centenary. Edition of 1921*, M. Lowe, R. S. Cohen and Y. Elkana eds and tr., Dordrecht, D. Reidel Publishing Company.

Hilbert D. 1899. *Grundlagen der Geometrie*, Leipzig, Teubner.

Hilbert D. 1900. Über den Zahlbegriff, *Jahrbericht der Deutschen Mathematiker-Vereinigung*, 8, pp. 180–84.

Hilbert D. 1904. Foundations of Logic and Arithmetic, in Heijenhoort (1963), pp. 129–38.

Hilbert D. 1922. Neubegründung der Mathematik, *Abhandlungen aus dem Mathematischen Seminar der Universität Hamburg* , Bd I, pp. 155–77.

Hintikka J. 1988. On the Development of the Model-Theoretic Viewpoint in Logical Theory, *Synthese*, 77: 1–36.

Hodges W. 1993. *Model Theory*, Cambridge, Cambridge University Press.

Hölder O. 1901. Die Axiome der Quantität und die Lehre vom Mass, *Berichte über die Verhandlungen der Königlich Sächsischen Gesellschaft der Wissenschaften zu Leipzig, Mathematisch-Physikaliche Classe*, 53, pp. 1–64. (English translation by J. Michell and C. Ernst, *Journal of Mathematical Psychology*, 40, pp. 235–52, 1996 (part I); 41, pp. 345–56, 1997 (part II)).

Hollings C. 2009. The Early Development of the Algebraic Theory of Semigroups, *Archive for History of Exact Sciences*, 63(5): 497–536.

Howie J. M. 1995. *Fundamentals of Semigroup Theory*, Oxford, Clarendon Press.

Huntington E. V. 1902. A Complete Set of Postulates for the Theory of Absolute Continuous Magnitudes, *Transactions of the American Mathematical Society*, 3: 264–79.

Hylton P. 1989. The Significance of *On Denoting*, in Anderson and Savage, pp. 88–107.

Hylton P. 1990. *Russell, Idealism and the Emergence of Analytical Philosophy*, Oxford, Clarendon Press.

Irvine A. D. 1989. Epistemic Logicism and Russell's Regressive Method, *Philosophical Studies*, 55(3): 303–27.

Johnson W. E. 1921. *Logic*, Part I, Cambridge, Cambridge University Press.

Johnson W. E. 1922. *Logic*, Part II, Cambridge, Cambridge University Press.

Johnson W. E. 1924. *Logic*, Part III, Cambridge, Cambridge University Press.

Jordan C. 1882. *Cours d'Analyse*, Paris, Gauthier-Villars.

Kant I. 1781. *Critique of Pure Reason*, English tr. by P. Guyer and A. Wood, Cambridge, Cambridge University Press, 1998.

Kant I. 1786. *Metaphysical Foundations of Natural Science*, English tr. by M. Friedman, Cambridge, Cambridge University Press, 2004.

Klein F. 1871. Ueber die sogenannte Nicht-Euklidische Geometrie, *Mathematische Annalen*, 4, pp. 573–675. (English tr. in Stillwell (1996), pp. 69–112.)

Klein F. 1873. Ueber die sogenannte Nicht-Euklidische Geometrie, zweiter Aufsatz, *Mathematische Annalen*, 6: 311–52.

Klein F. 1872. Vergleichende Betrachtungen über neuere geometrische Forschungen, *Mathematische Annalen*, 43(1893): 63–100.

Klein F. 1925. *Elementary Mathematics from an Advanced Standpoint*, English tr. by E. Hedrick and C. Nobles, Mineola, Dover.

Klein F. 1928. *Vorlesungen über nicht-euklidische Geometrie*, Berlin, J. Springer.

Krantz D. H., Luce R. D., Suppes P. and Tversky A. 1971. *Foundations of Measurement, vol. I: Additive and Polynomial Representations*, New York, Academic Press.

Landini G. 1998. *Russell's Hidden Substitutional Theory*, Oxford, Oxford University Press.

Landini G. 2006. Fregean Cardinals as Concept-Correlates, *Erkenntniss*, 65: 207–43.

Landini G. 2007. *Wittgenstein's Apprenticeship with Russell*, Cambridge, Cambridge University Press.

Langford C. H. 1942. The Notion of Analysis in Moore's Philosophy, in M. Schlipp ed., *The Philosophy of G.-E. Moore*, Evanston and Chicago, Northwestern University Press, pp. 321–42.

Laplace P. S. 1808. *Exposition du système du monde*, 3rd edn, Paris, Courcier.

Larvor B. 2010. Review of Paolo Mancosu, *The Philosophy of Mathematical Practice*, *Philosophia Mathematica*, 18(3): 350–60.

Lebesgue H. 1935. Sur la mesure des grandeurs, *L'enseignement mathématique*, vols 31–34 (Repr., Paris, Blanchard, 1975).

Leibniz G. W. 1684. *Meditations on Knowledge, Truth, and Ideas*, tr. by L. E. Loemker, *Philosophical Papers and Letters: A Selection*, Dordrecht, Boston, London, Kluwer, Leibniz G. W. 1975, pp. 291–96.

Levine J. 1998. From Absolute Idealism to *The Principles of Mathematics*, Critical Notice of *The Collected Papers of B. Russell* vols. 2–3, *International Journal of Philosophical Studies*, 6(1): 87–127.

Levine J. 2001. Logical Form, General Sentences, and Russell's Path to 'On Denoting', in Richard Gaskin ed., *Grammar in Early Twentieth-Century Philosophy*, London, Routledge, pp. 74–115.

Levine J. 2002. Analysis and Decomposition in Frege and Russell, *The Philosophical Quarterly*, 52(207): 195–216.

Levine J. 2009. From Moore to Peano to Watson. The Mathematical Roots of Russell's Naturalism and Behaviorism, *The Baltic International Yearbook of Cognition, Logic and Communication*, http://thebalticyearbook.org/, Volume 4: 200 Years of Analytical Philosophy, pp. 1–126.

Lindenbaum A. 1935. Sur la simplicité formelle des notions, *Actes du congrès international de philosophie scientifique*, Hermann, Paris, pp. 29–38.

Linsky B. 1999. *Russell's Metaphysical Logic*, Stanford, CSLI Publication.

Löwe B. and Müller T. 2008. Mathematical Knowledge is Context-Dependent, *Grazer Philosophische Studien*, 76: 91–107.

Makin G. 2000. *The Metaphysicians of Meaning: Russell and Frege on Sense and Denotation*, New York, Routledge.

Mancosu P. 2008. *The Philosophy of Mathematical Practice*, Oxford, Oxford University Press.

Mancosu P., Jorgensen K. F. and Pedersen S. A. 2005. *Visualization, Explanation and Reasoning Styles in Mathematics*, Heidelberg and New York, Springer-Verlag.

Manders K. 1987. Logic and Conceptual Relationships in Mathematics, in *Logic Colloquium '85*, The Paris Logic Group ed., Amsterdam, Elsevier, pp. 193–212.

Manders K. 1989. Domain Extensions and the Philosophy of Mathematics, *Journal of Philosophy*, 86: 553–62.

Marchisotto E. and Smith J. T. 2007. *The Legacy of Mario Pieri in Geometry and Arithmetic*, Boston, Birkhäuser.

Marion M. 1998. *Wittgenstein, Finitism and the Foundations of Mathematics*, Oxford, Oxford University Press.

Marion M. 2004. *Wittgenstein, Introduction au* Tractatus Logico Philosophicus, Paris, PUF.

Meinong A. 1896. *Über die Bedeutung des Weber'schen Gesetzes*, Leipzig, L. Voss (repr. in *Meinong's Gesammelte Abhandlungen*, Vol. 2, Leipzig, J. A. Barth, 1913).

Meinong A. 1902. *Über Annahme*, Leipzig, J. A. Barth.

Michel A. 1992. *Constitution de la théorie moderne de l'intégration*, Paris, Vrin.

Michell J. 1997. Bertrand Russell's 1897 Critique of the Traditional Theory of Measurement, *Synthese*, 110: 257–76.

Michell J. 1999. *Measurement in Psychology. A Critical History of a Methodological Concept*, Cambridge, Cambridge University Press.

Molk J. ed. 1904–16. *Encyclopédie des sciences mathématiques pures et appliquées publiées sous les auspices des académies des sciences de Göttingue, de Leipzig, de Munich et de Vienne avec la collaboration de nombreux savants*, Paris, Gauthier-Villars (repr. Paris, Gabay, 1991).

Moore G. E. 1899. The Nature of Judgement, *Mind*, 8: 167–93.

Moore G. E. 1903. *Principia Ethica*, Cambridge, Cambridge University Press.

Moore G. H. 1982. *Zermelo's Axiom of Choice: Its Origins, Development, and Influence*, New York-Heidelberg-Berlin, Springer-Verlag.

Morgan, A. de 1849. *Trigonometry and Double Algebra*, London, Taylor, Walton and Maberly.

Mühlhölzer F. 2005. 'A Mathematical Proof Must Be Surveyable': What Wittgenstein Meant By This and What It Implies, *Grazer Philosophische Studien*, 71: 57–86.

Mundy B. 1987. The Metaphysics of Quantity, *Philosophical Studies*, 51: 29–54.

Musgrave A. 1977. Logicism Revisited, *British Journal of Philosophy of Science*, 38: 99–127.

Nabonnand P. 2000a. La polémique entre Poincaré et Russell au sujet du statut des axiomes de la géométrie, *Revue d'Histoire des Mathématiques*, 6(2): 219–69.

Nabonnand P. 2000b. La genèse psycho-physiologique de la géométrie selon Poincaré, *Textes du séminaire 'Histoires de Géométrie'*, Fondation Maison des Sciences de l'Homme, Paris.

Nabonnand P. 2008. La théorie des *Würfe* de von Staudt – Une irruption de l'algèbre dans la géométrie pure, *Archive for History of Exact Sciences*, 62: 201–42.

Pasch M. 1882. *Vorlesungen über neure Geometrie*, Leipzig, Teubner.

Paseau A. 2005. What the Foundationalist Filter Kept Out, *Studies in History and Philosophy of Science*, 36: 191–201.

Peano G. 1887. *Applicazioni geometriche del calcolo infinitesimale*, Torino, Fratelli Bocca.

Peano G. 1888. *Calcolo Geometrico secundo l'Ausdehnungslehre di H. Grassmann*, Torino, Fratelli Bocca (English tr. by L. Kannenberg, Basel, Birkhauser, 2002).

Peano G. 1889a. Arithmetices principia, nova methodo exposita, Torino, Bocca (English tr. in Kennedy (1973), pp. 101–34).

Peano G. 1889b. I principii di geometria logicamente esposti, Torino, Bocca.

254 *Bibliography*

Peano G. 1894. Notations de Logique Mathématiques. Introduction au Formulaire de mathématiques, *Rivista di Matematica*, IV.

Peano G. 1896. Saggio di calcolo geometrico, *Atti della Reale Accademia delle Scienze di Torino*, 31: 952–75.

Peano G. 1899. Sui numeri irrazionali *Rivista di Matematica*, IV: 126–40.

Peano G. 1903. La geometria basata sulle idee di punto e distanza, *Atti della R. Accademia delle Scienze di Torino*, pp. 6–10.

Pears D. F. 1967. *Bertrand Russell and the British Tradition in Philosophy*, London, Collins.

Pieri M. 1898. I principii della geometria di posizione composti in sistema logico deduttivo, *Memoria della R. Accademia delle scienze di Torino*, 48: 1–62.

Pieri M. 1900. Della geometria elementare como sistema ipotetico deduttivo: Monografia del punto e del moto, *Memorie della Reale Accademia delle Scienze di Torino* (series 2), 49: 173–222.

Pieri M. 1908. La geometria elementare istituita sulle nozioni di 'punto' e 'sfera', *Memorie di matematica e di fisica della Società Italiana delle Scienze* (series 3) 15: 345–450 (Tr. In Marchisotto and Smith (2007), pp. 160–270).

Plotkin J. L. ed. 2005. *Hausdorff on Ordered Sets*, Providence, American Mathematical Society.

Poincaré H. 1893. Le continu mathématique, *Revue de Métaphysique et de Morale*, 1: 26–34.

Poincaré H. 1898. On the Foundations of Geometry, *The Monist*, 9: 1–43.

Poincaré H. 1899. Des fondements de la géométrie: à propos d'un livre de M. Russell, *Revue de Métaphysique et de Morale*, 7: 251–79.

Poincaré H. 1902. *La Science et l'Hypothèse*, Paris, Flammarion (English tr., London and Newcastle-on-Tyne, Walter Scott Publishing, 1905).

Poincaré H. 1905. *La Valeur de la Science*, Paris, Flammarion (English tr. by B. Halsted (1913), New York, Cosimo (2007)).

Poincaré H. 1908. *Science et Méthode*, Paris, Flammarion (English tr. by F. Maitland (1914), Mineola, Dover (2003)).

Poincaré H. 1913. *Dernières pensées*, Paris, Flammarion.

Poincaré H. 2002. *L'opportunisme scientifique*, Basel, Birkhäuser.

Poli R. 2004. W. E. Johnson's Determinable-Determinate Opposition and his Theory of Abstraction, in *Idealization XI: Historical Studies on Abstraction and Idealization, Poznan Studies in the Philosophy of the Sciences and the Humanities*, 82: 163–96.

Potrc M. and Vospernik M. 1996. Meinong on Psychological Measurement, *Axiomathes*, 1–2: 187–202.

Potter M. and Ricketts T. 2010. *The Cambridge Companion to G. Frege*, Cambridge, Cambridge University Press.

Prior A. 1949. Determinables, Determinates and Determinants, *Mind*, 58(229): 1–30; 58(230): 178–94.

Quine W. V. O. 1941. Whitehead and the Rise of Modern Logic, in *Philosophy of A. N. Whitehead*, P. A. Schilpp ed., LaSalle, Open Court, pp. 125–63 (quoted from *Selected Logic Papers*, Cambridge (MA), Harvard University Press, 1995, pp. 3–36).

Quine W. V. O. 1953. *From a Logical Point of View*, Harvard, Harvard University Press, (quoted from the 2nd edn, 1994).

Quine W. V. O. 1963. *Set Theory and Its Logic*, Cambridge (MA), Belknap Press.

Ramsey F. P. 1929. *Theories*, in Ramsey (1990), pp. 112–36.

Ramsey F. P. 1990. *Philosophical Papers*, ed. D. H. Mellor, Cambridge, Cambridge University Press.

Rawls J. 1971. *A Theory of Justice*, Cambridge (MA), Belnap Press of Harvard.

Rayo A. 2005. Logicism Reconsidered, in S. Shapiro ed., *The Oxford Handbook of Philosophy of Mathematics and Logic*, Oxford, Oxford University Press, pp. 203–35.

Richards J. L. 1988. *Mathematical Visions: The Pursuit of Geometry in Victorian England*, Boston, Academic Press.

Rivenc F. 1993. *Recherches sur l'universalisme logique*, Paris, Payot.

Rodriguez-Consuegra F. 1991. *The Mathematical Philosophy of Bertrand Russell: Origins and Development*, Basel, Birkhäuser.

Rouilhan (de) P. 1996. *Russell et le cercle des paradoxes*, Paris, PUF.

Russell, B. 1983–in progress. *The Collected Papers of Bertrand Russell*, 36 volumes planned, 17 already published, London and New York, Routledge.

Russell, B. 1896–98. Various Notes on Mathematical Philosophy, *The Collected Papers of Bertrand Russell, Vol. 2*, pp. 6–29.

Russell, B. 1897a. *An Essay on the Foundations of Geometry*, Cambridge, Cambridge University Press (repr. London and New York, Routledge, 1996).

Russell, B. 1897b. Review of Couturat, *De l'Infini mathématique*, Mind, 6: 112–19 (*The Collected Papers of Bertrand Russell, Vol. 2*, pp. 60–7).

Russell, B. 1897c. On the Relations of Numbers and Quantity, *Mind*, 6: 326–41 (*The Collected Papers of Bertrand Russell, Vol. 2*, pp. 70–82).

Russell, B. 1898a. On Quantity and Allied Conceptions, *The Collected Papers of Bertrand Russell, Vol. 2*, pp. 114–35.

Russell, B. 1898b. An Analysis of Mathematical Reasoning Being an Inquiry into the Subject-Matter, the Fundamental Conceptions, and the Necessary Postulates of Mathematics, *The Collected Papers of Bertrand Russell, Vol. 2*, pp. 155–242.

Russell, B. 1898c. On the Constituents of Space and Their Mutual Relations, *The Collected Papers of Bertrand Russell, Vol. 2*, pp. 309–21.

Russell, B. 1898d. Note on Order, *The Collected Papers of Bertrand Russell, Vol. 2*, pp. 339–58.

Russell, B. 1899a. The Classification of Relations, *The Collected Papers of Bertrand Russell, Vol. 2*, pp. 138–46.

Russell, B. 1899b. Notes on Geometry, *The Collected Papers of Bertrand Russell, Vol. 2*, pp. 359–89.

Russell, B. 1899c. Miscellaneous Notes: 'Fragments on Series', *The Collected Papers of Bertrand Russell, Vol. 2*, pp. 457–59.

Russell, B. 1899d. Review of Meinong's *Ueber die Bedeutung des Weber'schen Gesetzes*, *Mind*, 8, pp. 251–56 (*The Collected Papers of Bertrand Russell, Vol. 2*, pp. 148–52).

Russell, B. 1899e. Sur les axiomes de la géométrie, *Revue de Métaphysique et de Morale*, 7, pp. 684–707.

Russell, B. 1899f. The Fundamental Ideas and Axioms of Mathematics, *The Collected Papers of Bertrand Russell, Vol. 2*, pp. 245–305.

Russell, B. 1900a. *A Critical Exposition of the Philosophy of Leibniz with an Appendix of Leading Passages*, Cambridge, Cambridge University Press.

Russell, B. 1900b. On the Logic of Relations with Applications to Arithmetic and the Theory of Series, *The Collected Papers of Bertrand Russell, Vol. 3*, pp. 590–612.

Russell, B. 1900c. Do Differences Differ?, *The Collected Papers of Bertrand Russell, Vol. 3*, pp. 556–57.

Russell, B. 1901a. Sur la logique des relations avec des applications à la théorie des séries, *Rivista di Matematica*, 7: 115–36, 137–48 (*The Collected Papers of Bertrand Russell, Vol. 3*, pp. 613–27).

Russell, B. 1901b. Recent Italian Works on the Foundations of Mathematics, *International Monthly*, 4: 83–101 (*The Collected Papers of Bertrand Russell, Vol. 3*, pp. 352–62).

Russell, B. 1901c. Is Position in Time and Space Relative or Absolute? *Mind*, 10(30): 293–317 (*The Collected Papers of Bertrand Russell, Vol. 3*, pp. 259–84).

Russell, B. 1902. Geometry, Non-Euclidean. *Encyclopaedia Britannica*, 28, pp. 664–74 (*The Collected Papers of Bertrand Russell, Vol. 3*, pp. 474–504).

Russell, B. 1903. *The Principles of Mathematics*, Cambridge, Cambridge University Press.

Russell, B. 1904. Meinong's Theory of Complexes and Assumptions, *Mind*, 13: 204–19, 336–54, 509–24 (*The Collected Papers of Bertrand Russell, Vol. 4*, pp. 432–74).

Russell, B. 1905a. Review of Poincaré, *Science and Hypothesis*, *Mind*, 14: 412–18 (*The Collected Papers of Bertrand Russell, Vol. 4*, pp. 589–94).

Russell, B. 1905b. Sur la Relation des Mathématiques à la Logistique, *Revue de Métaphysique et de Morale*, 13: 906–17 (English tr. in Russell (1973), pp. 260–71).

Russell, B. 1905c. On Denoting, *Mind*, 14(4): 479–93.

Russell, B. 1907a. The Study of Mathematics, *New Quarterly*, 1: 29–44 (*The Collected Papers of Bertrand Russell, Vol. 4*, pp. 85–93).

Russell, B. 1907b. The Regressive Method of Discovering the Premises of Mathematics, in Russell (1973), pp. 272–82.

Russell, B. 1912a. *The Problems of Philosophy*, London: Williams and Norgate/ New York, H. Holt (repr. Oxford, Oxford University Press, 1972).

Russell, B. 1912b. What is Logic? *The Collected Papers of Bertrand Russell, Vol. 6*, pp. 54–6.

Russell, B. 1913. *The Theory of Knowledge, The Collected Papers of Bertrand Russell, Vol. 7*.

Russell, B. 1914a. The Relation of Sense Data to Physics, *Scientia*, 4 (*The Collected Papers of Bertrand Russell, Vol. 8*, pp. 3–26).

Russell, B. 1914b. *Our Knowledge of the External World as a Field for Scientific Method in Philosophy*, London, Open Court.

Russell, B. 1919. *Introduction to Mathematical Philosophy*, London, Allen and Unwin.

Russell, B. 1924. Logical Atomism, in J. H. Muirhead ed., *Contemporary British Philosophy: Personal Statements*, pp. 357–83 (*The Collected Papers of Bertrand Russell, Vol. 9*, pp. 162–79).

Russell, B. 1927. *Analysis of Matter*, London: Kegan Paul.

Russell, B. 1937. *The Principles of Mathematics*, 2nd edn, Padstow, T. J. Press Ltd.

Russell, B. 1959. *My Philosophical Development*, London, Allen & Unwin.

Russell, B. 1969. *The Autobiography of Bertand Russell*, 3 vols, London, Allen & Unwin.

Russell, B. 1973. *Essays in Analysis*, ed. D. Lackey, London, Allen & Unwin.

Russell B. and Whitehead A. N. 1910. *Principia Mathematica, Vol. I*, Cambridge, Cambridge University Press.

Russell B. and Whitehead A. N. 1912. *Principia Mathematica, Vol. II*, Cambridge, Cambridge University Press.

Russell B. and Whitehead A. N. 1913. *Principia Mathematica, Vol. III*, Cambridge, Cambridge University Press.

Sackur J. 2005. *Formes et faits. Analyse et théorie de la connaissance dans l'atomisme logique*, Paris, Vrin.

Schmid A.-F. ed. 2001. *Bertrand Russell. Correspondance sur la philosophie, la logique et la politique avec Louis Couturat (1897–1913)*, Vols I–II, Paris, Kimé.

Séguier (de) J. A. M. J. 1904. *Théorie des groupes finis. Éléments de la théorie des groupes abstraits*, Paris, Gauthier-Villars.

Shapiro S. 1997. *Philosophy of Mathematics: Structure and Ontology*, Oxford, Oxford University Press.

Shapiro S. 2000. Frege Meets Dedekind: A Neologicist Treatment of Real Analysis, *Notre Dame Journal of Formal Logic*, 41(4): 335–64.

Shapiro S. 2006. Structure and Identity, in F. McBride ed., *Identity and Modality*, Oxford, Oxford University Press, pp. 109–45.

Shapiro S. and Weir A. 1999. New V, ZF and Abstraction, *Philosophia Mathematica*, 7: 293–321.

Simons P. M. 1987. Frege's Theory of Real Numbers, *History and Philosophy of Logic*, 8: 25–44.

Sinaceur H. 1991. *Corps et Modèles. Essai sur l'histoire de l'algèbre réelle*, Vrin, Paris.

Smith B. 1994. *Austrian Philosophy. The Legacy of Franz Brentano*, Chicago-La Salle, Open Court.

Steiner M. 1998. *The Applicability of Mathematics as a Philosophical Problem*, Cambridge (MA), Harvard University Press.

Stevens G. 2005. *The Russellian Origins of Analytic Philosophy*, London, Routledge.

Stevens S. S. 1946. On the Theory of Scales of Measurement, *Science*, 103: 667–80.

Stevens S. S. 1986. *Psychophysics – Introduction to its Perceptual, Neural and Social Prospects*, Piscataway, Transaction Publishers.

Stillwell J. 1996. *Sources of Hyperbolic Geometry*, Providence, American Mathematical Society.

Stolz O. 1885. *Vorlesungen über allgemeine Arithmetik*, Leipzig, Teubner.

Struve H. and Struve R. 2004. Projective Spaces with Cayley–Klein Metrics, *Journal of Geometry*, 81: 155–67.

Suppes P. 1951. A Set of Independent Axioms for Extensive Quantities, *Portugaliae Mathematica*, 10: 163–72.

Suppes P. and Zinnes J. 1968. Basic Measurement Theory, in Luce, R. D. *et al.* eds, *Handbook of Mathematical Psychology, Vol. I*, New York, John Wiley & Sons.

Suschkewitsch A. 1928. Über die endlichen Gruppen ohne das Gesetz der eindeutigen Umkehrbarkeit, *Mathematische Annalen*, 99(1): 30–50.

Tappenden J. 2008. Mathematical Concepts: Fruitfulness and Naturalness, in Mancosu (2008), pp. 276–301.

Torretti R. 1978. *Philosophy of Geometry from Riemann to Poincaré*, Dordrecht, Reidel.

Van Heijenoort J. ed. 1963. *From Frege to Gödel. A Source Book in Mathematical Logic, 1879–1931*, Harvard University Press.

Van Kerkhove B. ed. 2009. *New Perspectives on Mathematical Practices*, Singapore, World Scientific Publishing.

Veblen O. 1906. The Square Root and the Relation of Order, *Transactions of the American Mathematical Society*, 7: 197–99.

Veblen O. and Bussey W. H. 1906. Finite Projective Geometries, *Transactions of the American Mathematical Society*, 7: 241–59.

Veblen O. and Young J. W. 1910. *Projective Geometry, Vol. I*, Boston, Ginn & Co.

Veblen O. and Young J. W.1914. *Projective Geometry, Vol. II*, Boston, Ginn & Co.

Veronese G. 1891. *Fondamenti di geometria a più dimensioni e a più specie di unità rettilinee esposti in forma elementare*, Lezioni per la Scuola di magistero in Matematica, Padova, Tipografia del Seminario.

Voelke J.-D. 2005. *Renaissance de la géométrie non euclidienne entre 1865 et 1900*, Berne, Peter Lang.

Von Staudt G. 1847. *Geometrie der Lage*, Nürnberg, Verlag von Bauer and Raspe. (Italian tr. by M. Pieri, *Geometria di posizione di Georgio von Staudt*, Torino, Bocca, 1889.)

Wahl R. 2007. 'On Denoting' and the Principle of Acquaintance, *Russell: The Journal of Bertrand Russell Studies*, 27: 7–23.

Walsh S. 2010. Logicism, Interpretability and Knowledge of Arithmetic, www.nd.edu/~swalsh/papers/paper-interpretability-draft-04-07-2010.pdf.

Weber H. 1895–96. *Lehrbuch der Algebra, Erster Band*, Braunschweig, F. Vieweg und Sohn.

Weil A. 1978. Who Betrayed Euclid?, *Archive for History of Exact Sciences*, 19: 91–3.

Weiner J. 1990. *Frege in Perspective*, London, Cambridge University Press.

Weiner J. 2007. What's in a Numeral? Frege's answer, *Mind*, 116(463): 677–716.

White M. J. 1992. *The Continuous and the Discrete: Ancient Physical Theories from a Contemporary Perspective*, Oxford, Clarendon Press.

Whitehead A. N. 1898. *A Treatise on Universal Algebra with Applications*, Cambridge, Cambridge University Press (repr. New York, Hafner, 1960).

Whitehead A. N. 1906a. *On Mathematical Concepts of the Material World*, London, Dulau.

Whitehead A. N. 1906b. *The Axioms of Projective Geometry*, Cambridge, Cambridge University Press (repr. New York, Hafner, 1971).

Whitehead A. N. 1907. *The Axioms of Descriptive Geometry*, Cambridge, Cambridge University Press (repr. New York, Hafner, 1971).

Whitehead A. N. 1911. *An Introduction to Mathematics*, London, Williams & Northgate.

Whitehead A. N. 1916. La théorie relationniste de l'espace, *Revue de Métaphysique et de Morale*, 23: 423–54.

Whitehead A. N. 1919. *An Enquiry concerning the Principles of Natural Knowledge*, Cambridge, Cambridge University Press.

Whitehead A. N. 1920. *The Concept of Nature*, Cambridge, Cambridge University Press.

Whitehead A. N. 1922. *The Principle of Relativity with Applications to Physical Science*, Cambridge, Cambridge University Press.

Whitehead A. N. 1925. *Science and the Modern World*, New York, Free Press.

Whitehead A. N. 1928. *Process and Reality*, Gifford Lectures Delivered in the University of Edinburgh during the Session 1927–28, New York, Free Press.

Wiener N. 1976. *Collected Works of N. Wiener*, vol. I, Cambridge (MA), MIT Press.

Wiener N. 1921. A New Theory of Measurement: A Study in the Logic of Mathematics, *Proceedings of the London Mathematical Society*, 19: 185–205.

Wigner E. 1960. The Unreasonable Effectiveness of Mathematics in the Natural Sciences, *Communications in Pure and Applied Mathematics*, 13(1): 1–14.

Wilson M. 1994. Can We Trust Logical Form?, *Journal of Philosophy*, 91: 519–44.

Wilson M. 2008. *Wandering Significance: An Essay on Conceptual Behaviour*, Oxford, Oxford University Press.

Wilson M. 2010. Frege's Mathematical Setting, in Potter and Ricketts (2010), pp. 379–412.

Wittgenstein L. 1921. *Logisch-Philosophische Abhandlung*, Wilhelm Ostwald ed., *Annalen der Naturphilosophie*, 14 (English tr. C. K. Ogden, London, Kegan Paul, 1922).

Wittgenstein L. 1929. Some Remarks on Logical Forms, *Proceedings of the Aristotelian Society*, Supplementary Volume, 9, pp. 162–71.

Wittgenstein L. 1953. *Philosophical Investigations*, Oxford, Blackwell.

Wittgenstein L. 1956. *Remarks on the Foundations of Mathematics*, Oxford, Blackwell.

Wittgenstein L. 1974. *Philosophical Grammar*, Oxford, Blackwell.

Wittgenstein L. 1979. *Ludwig Wittgenstein and the Vienna Circle: Conversations Recorded by Friedrich Waismann*, B.F. McGuinness ed., Oxford, Blackwell.

Wright C. 1983. *Frege's Conception of Numbers as Objects*, Aberdeen, Aberdeen University Press.

Wright C. 2000. Neo-Fregean Foundations for Real Analysis: Some Reflections on Frege's Constraint, *Notre Dame Journal of Formal Logic*, 41(4): 317–33.

Index

Printed in the United States
By Bookmasters